Great Power Cyber Competition

This volume conceptualizes the threats, challenges, opportunities, and boundaries of great power cyber competition of the 21st century.

This book focuses on a key dimension of contemporary great power competition that is often less understood due to its intangible character: the competition taking place in the cyber domain, including information and cyber operations. Democracies across the globe find themselves in an unrelenting competition with peer and near-peer competitors, with a prevailing notion that no state is "safe" from the informational contest. Adversarial powers, particularly China and Russia, recognize that most competition is principally non-kinetic but dominates the information environment and cyberspace, and the volume articulates the Russian and Chinese strategies to elevate cyber and information competition to a central position. Western governments and, in particular, the U.S. government have long conceived of a war–peace duality, but that perspective is giving way to a more nuanced perception of competition. This volume goes beyond analyzing the problems prevalent in the information space and offers a roadmap for Western powers to compete in and protect the global information environment from malicious actors. Its genesis is rooted in the proposition that it is time for the West to push back against aggression and that it needs a relevant framework and tools to do so. The book demonstrates that Western democratic states currently lack both the strategic and intellectual acumen to compete and win in the information and cyber domains, and argues that the West needs a strategy to compete with near-peer powers in information and cyber warfare.

This book will be of much interest to students of cyber-warfare, information warfare, defense studies, and international relations in general, as well as practitioners.

David V. Gioe is a British Academy Global professor and visiting professor of intelligence and international security in the Department of War Studies at King's College London. He is also an associate professor of history at the U.S. Military Academy and a history fellow for its Army Cyber Institute.

Margaret W. Smith is an active-duty cyber officer in the U.S. Army, a senior fellow with the Atlantic Council's Cyber Statecraft Initiative, and graduate faculty at the University of Maryland. She holds a PhD in Public Policy from The George Washington University.

Routledge Advances in Defence Studies

Routledge Advances in Defence Studies is a multi-disciplinary series examining innovations, disruptions, counter-culture histories, and unconventional approaches to understanding contemporary forms, challenges, logics, frameworks, and technologies of national defence. This is the first series explicitly dedicated to examining the impact of radical change on national security and the construction of theoretical and imagined disruptions to existing structures, practices, and behaviours in the defence community of practice. The purpose of this series is to establish a first-class intellectual home for conceptually challenging and empirically authoritative studies that offer insight, clarity, and sustained focus.

Series editors: Timothy Clack, *University of Oxford, UK*, and Oliver Lewis *Rebellion Defence* and *University of Southern California, USA*

Advisory Board: Tarak Barkawi *London School of Economics, UK* Richard Barrons *Global Strategy Forum, UK* Kari Bingen-Tytler *Center for Strategic and International Studies, USA* Ori Brafman *University of California, Berkeley, USA* Tom Copinger-Symes *British Army, UK* Karen Gibsen *Purdue University, USA* David Gioe *West Point, USA* Robert Johnson *Oxford University, UK* Mara Karlin *John Hopkins University, USA* Tony King *Warwick University, UK* Benedict Kite, *British Army, UK* Andrew Sharpe *Centre for Historical and Conflict Research, UK* Suzanne Raine *Cambridge University, UK*

Great Power Cyber Competition
Competing and Winning in the Information Environment
Edited by David V. Gioe and Margaret W. Smith

Human Factors in Effective Counter-Terrorism
A Comparative Study
Richard Warnes

For more information about this series, please visit: www.routledge.com/Routledge-Advances-in-Defence-Studies/book-series/RAIDS

Great Power Cyber Competition

Competing and Winning in the Information Environment

Edited by David V. Gioe and Margaret W. Smith

Routledge
Taylor & Francis Group

LONDON AND NEW YORK

First published 2024
by Routledge
4 Park Square, Milton Park, Abingdon, Oxon OX14 4RN

and by Routledge
605 Third Avenue, New York, NY 10158

Routledge is an imprint of the Taylor & Francis Group, an informa business

British Library Cataloguing-in-Publication Data
A catalogue record for this book is available from the British Library

ISBN: 978-1-032-54526-4 (hbk)
ISBN: 978-1-032-54529-5 (pbk)
ISBN: 978-1-003-42530-4 (ebk)

DOI: 10.4324/9781003425304

Typeset in Times New Roman
by Newgen Publishing UK

Contents

List of Contributors   *vii*
Acknowledgments   *xii*

Introduction   1
DAVID V. GIOE AND MARGARET W. SMITH

1   A Strategic Cyberspace Overview: Russia and China   6
    MARK GRZEGORZEWSKI AND CHRISTOPHER MARSH

2   On Competition: A Continuation of Policy by
    Misunderstood Means   25
    JAYSON WARREN

3   Russian New Generation Warfare in the Baltic States
    and Beyond   44
    SANDOR FABIAN AND JANIS BERZINS

4   Russian Cyberspace Operations against Ukraine in the 2022
    War: How Effective Have They Been and What Lessons for
    NATO Can Be Drawn?   57
    MARINA MIRON AND ROD THORNTON

5   Everyone a Sensor: The Implications of the Russo-Ukrainian
    War and the Democratization of Intelligence for Great Power
    Competition   73
    DAVID V. GIOE AND TONY MANGANELLO

6   In Africa, Great Power Competition Requires a
    Great Strategy for Information Operations   87
    TARA HEIDGER AND DAVID HIGGINS

7 Competing for Influence: Authoritarian Powers in the
Cyber Domain in Latin America 100
FABIANA SOFIA PERERA

8 The Logic of Protraction in Cyber Conflict: Peace
Would Ruin Me 115
TREY HERR, EMMA SCHROEDER, AND STEWART SCOTT

9 Digital IEDs on the Information Highway: PSYOPS,
CYBER, and the Info Fight 127
CHAVESO L. COOK

10 Cybersecurity as a Public Good: Government Intervention
Is Only Part of the Solution 142
MARGARET W. SMITH AND JIM MONKEN

11 Unconventional Warfare in the Information Environment 157
OTTO C. FIALA AND JIM WORRALL

12 Ubiquitous Technical Surveillance and the Challenges of
Military Operations in the Era of Great Power Competition 173
CHRISTOPHER CRUDEN

13 Toward a Whole-of-Society Framework for Countering
Disinformation 183
J.D. MADDOX, CASI GENTZEL, AND ADELA LEVIS

14 Enduring Challenges in Cybersecurity: Responding
Quickly and Credibly to Asymmetric Threats 199
MICHAEL POZNANSKY

Index *213*

Contributors

Janis Berzins is Director of the Center for Security and Strategic Research at the National Defense Academy of Latvia, a nonresident research fellow and senior advisor at the Swedish Defence University, and a senior research fellow at the Center for the Study of New Generation Warfare. Dr. Berzins' research focuses on the juxtaposition between the theoretical developments of Russian military thought and the operational reality on the ground.

Chaveso L. Cook (COL, United States Army) is a senior fellow with the Center for Junior Officers, a member of the Carnegie Council for Ethics in International Affairs, and a term member with the Council on Foreign Relations. He holds a PhD from Tufts University and master's degrees from Columbia University and the University of Texas at El Paso and is also the Cofounder and Executive Director of the nonprofit MilitaryMentors.org.

Christopher Cruden is a subject matter expert on the impact of technology on intelligence and military operations. He currently serves as an advisor to the U.S. government and private sector on matters related to intelligence and strategic policy. In this role, he provides guidance and implementation plans to senior staff on a wide range of crucial technology and tradecraft issues. In addition, Chris is a nonresident fellow at Joint Special Operations University.

Sandor Fabian is a former Hungarian Special Forces lieutenant colonel with more than 20 years of military experience. Dr. Fabian is a research associate at the University of Central Florida and a curriculum developer and team leader at LEIDOS. Dr. Fabian's research has appeared in the *Journal of Strategic Security, Defence Studies, Defense & Security Analysis, Special Operations Journal, Combating Terrorism Exchange, Florida Political Chronicle*, and *the Hungarian Seregszemle Journal*.

Otto C. Fiala, is contracted to NATO's Allied Command Transformation (ACT) as a Strategic Thinker by Dean Franks LLC. He was previously team lead for Sensitive Activities Research and Development at United States Army Special Operations Command and was also a resistance and resilience planner at SOCEUR, where he was also the chief editor and writer of the *Resistance Operating Concept*. He is a retired USAR civil affairs colonel.

Casi Gentzel is a foreign affairs officer with the U.S. Department of State's Global Engagement Center (GEC) who currently serves as the GEC's representative to the U.S. Indo-Pacific Command. Prior to joining the GEC in 2016, she served in the State Department's Bureau of Diplomatic Security and the Office of the Under Secretary for Civilian Security, Democracy, and Human Rights.

David V. Gioe is a British Academy Global professor and visiting professor of intelligence and international security in the Department of War Studies at King's College London. He is also an associate professor of history at the U.S. Military Academy and a history fellow for its Army Cyber Institute. Dr. Gioe is Director of Studies for the Cambridge Security Initiative and is Co-convener of its International Security and Intelligence Program. His analysis does not necessarily reflect any position of the U.S. government or Defense Department.

Mark Grzegorzewski is an assistant professor of cybersecurity at Embry Riddle University in the Security Studies and International Affairs Department. He is also an Army Cyber Institute fellow, *Special Operations Journal* book review editor, and is affiliated with the Joint Special Operations University, Irregular Warfare Center, and the National Intelligence University. He holds a PhD, MA, and BA in Political Science/Government from the University of South Florida, along with a graduate certificate in globalization studies.

Tara Heidger is a current Special Operations Command-Africa reservist and former active-duty psychological operations senior noncommissioned officer. As a civilian she currently advises U.S. information and influence strategy in Africa. Tara is a recent graduate of Columbia University, where she earned a dual master's degree in international affairs and urban planning, studying the intersection of East African development and conflict.

Trey Herr, is the Director of the Atlantic Council's Cyber Statecraft Initiative under the Digital Forensic Research Lab (DFRLab) and Assistant Professor of Cybersecurity and Policy at American University's School of International Service. At the Council his team works on cybersecurity and geopolitics including cloud computing, the security of the internet, supply chain policy, cyber effects on the battlefield, and growing a more capable cybersecurity policy workforce.

David Higgins works for a strategic communications agency and has previously worked for the UK government and United Nations in Somalia, East Africa, Afghanistan, and Iraq. David started his career as an active-duty infantry officer and is currently a lieutenant colonel in the British Army Reserve.

Adela Levis is a civil service officer with the U.S. Department of State's Global Engagement Center (GEC) and serves as the GEC Academic and Think-Tank Liaison. Prior to joining the GEC in 2015, she served in the State Department's Bureau for Democracy, Human Rights and Labor, and in the Bureau for Educational and Cultural Affairs.

J.D. Maddox supports the U.S. Department of State's GEC as CEO of Inventive Insights LLC. Maddox previously served as Deputy Coordinator of the GEC, as a branch chief in the Central Intelligence Agency, and as a U.S. Army Psychological Operations Team Leader. Maddox is also an adjunct professor in George Mason University's Department of Information Sciences and Technology.

Tony Manganello is an assistant professor of National Security Studies at Indiana Wesleyan University where he serves as Department Coordinator for the criminal justice program. He earned a PhD in War Studies from King's College London, and his research focuses on a range of intelligence and security topics. Prior to entering the world of higher education, Tony served 11 years as a special agent with the U.S. Secret Service.

Christopher Marsh, is Director of Research and Analysis in the Institute for SOF Strategic Studies at the Joint Special Operations University, U.S. Special Operations Command. He is the author of several books and dozens of articles on Russian and Chinese domestic, foreign, and defense policies, including *Unparalleled Reforms: China's Rise, Russia's Fall*, and the *Interdependence of Transition*. He is currently writing a book on great power competition between the United States, Russia, and China.

Marina Miron is working on a British Academy-funded project dealing with the threat of information war emanating from Russia and China and how to counter the current techniques and tools used by those actors, including AI and cyber "weapons." In addition to the information war, she researches Russian military doctrinal thinking, including doctrines for different branches of the Armed Forces and available and prospective weapons systems, with a particular focus on space assets and electronic warfare systems. Further, Dr. Miron has been working on several UK MoD projects related to human augmentation in the UK Armed Forces and assessing an ethical framework for introducing such technologies in a contemporary defense environment.

Jim Monken has spent nearly 20 years working within the national security, emergency preparedness, risk management, and energy resilience planning spheres. Throughout the past decade he has pioneered programs for critical information sharing, public and private sector integration, and large-scale exercise development and execution. He is also a 2002 graduate of United States Military Academy.

Fabiana Sofia Perera is an associate professor at the William J. Perry Center for Hemispheric Defense Studies and a non-resident fellow at the Modern War Institute. Prior to joining the Perry Center, Fabiana was a Rosenthal fellow at the Office of the Secretary of Defense, Under Secretary for Policy, Western Hemisphere Affairs. Fabiana holds an MA in Latin American Studies from Georgetown University and earned a PhD in Political Science from The George Washington University. Her research and analysis have appeared in numerous publications including *The Washington Post*, *CNN.com*, and *War on the Rocks*.

Michael Poznansky is an associate professor in the Strategic and Operational Research Department and a core faculty member in the Cyber and Innovation Policy Institute at the U.S. Naval War College. He previously taught in the Graduate School of Public and International Affairs at the University of Pittsburgh. His book, *In the Shadow of International Law: Secrecy and Regime Change in the Postwar World*, was published by Oxford University Press in 2020. Dr. Poznansky has held multiple fellowships and is the author of numerous articles in prominent *IR* and *Political Science* journals.

Emma Schroeder is an associate director with the Atlantic Council's Cyber Statecraft Initiative within the Digital Forensic Research Lab (DFRLab). She works primarily on CSI's Conflict, Tech, and Markets portfolio, which examines the role of cyber and cyber-enabled technology in conflict and its impact on states, non-state groups, and markets. Within this portfolio of work are three central topics: Conflict in and through Cyberspace, the Proliferation of Offensive Cyber Capabilities, and Combating Cybercrime.

Stewart Scott is an associate director with the Atlantic Council's Cyber Statecraft Initiative under the Digital Forensic Research Lab (DFRLab). He works on the Initiative's systems security portfolio, which focuses on software supply chain risk management and open-source software security policy.

Margaret W. Smith, is an active-duty cyber officer in the U.S. Army, a senior fellow with the Atlantic Council's Cyber Statecraft Initiative, and graduate faculty at the University of Maryland. Maggie's research focuses on information and cyber operations, the role of third-party actors in cyberspace, and the role of cyber during conflict. She holds an MPP in intelligence and homeland security policy from Georgetown University and a PhD in Public Policy from The George Washington University. Her research and analysis have appeared in numerous publications including *Lawfare*, the *Modern War Institute at West Point, Public Administration Review, Political Violence*, and *War on the Rocks*.

Rod Thornton is an associate professor in the Defence Studies Department of King's College London. Prior to academia, he spent nine years in a British Army infantry regiment during the Cold War. On leaving the Army, he went to university to study Russian and Serbo-Croat. After living for a while in both Kiev (now Kyiv) and Moscow, he subsequently went on to undertake a PhD. He has worked at several universities, including two in the Middle East. His research interests are in the general realm of security studies, looking mostly at the Russian armed forces. His publications include a book, *Asymmetric Warfare* (2008), and many articles on the Russian military. Most recently, his research has focused on Russian military high-tech systems – including cyber and AI – and how they are linked to the conduct of Information Warfare. He is currently writing a paper, with Dr. Marina Miron, on the Russian space program.

Jayson Warren is an active-duty military intelligence officer in the U.S. Air Force and PhD (Policy Studies) student at Clemson University. In addition to his graduate studies, he serves as the lead political scientist at Clemson University's Media Forensics Hub – an interdisciplinary team analyzing foreign-sponsored disinformation/coordinated inauthenticity campaigns on social media. Jayson holds an AA (equivalent) in Religion from Elet Szava Alapitvany (Hungary); a BA in Government and International Relations; and a MA in Public Policy from Liberty University where he is also a guest professor of Government and Strategic Intelligence.

Jim Worrall is a retired U.S. Army Special Forces lieutenant colonel with 30 years in the U.S. Army, 21 of those working in SOF. Since retiring, Jim has spent the last six years working in the EUCOM AOR. He is a graduate of the Chilean War College and holds a master's degree in international relations from the University of Miami.

Acknowledgments

The editors wish to thank Dr. Barnett Koven for his enthusiastic work in setting the foundations of this book and Dr. Marina Miron for her hard work and patience in getting the manuscript ready for publication. We also thank John Amble and the Modern Warfare Institute at West Point for their permission to expand on the initial blog project entitled, *Full-Spectrum: Capabilities and Authorities in Cyber and the Information Environment*. We thank the leaders of the Army Cyber Institute at West Point for their continued support of our research into the intersection of influence operations and the cyber domain. Finally, we gratefully acknowledge our chapter authors whose scholarship, good cheer, patience, and flexibility enabled us to put this volume together.

Additionally, David Gioe wishes to thank the British Academy's Global Professorship program for funding to work on disinformation while living in the United Kingdom and King's College London Department of War Studies for being a gracious and collegial host institution for the duration of the British Academy award.

Disclaimer: Many of our chapters' authors require disclaimers for their contributions due to their employer requirements. We thus offer that the analysis and views expressed in these chapters are those of the author(s) and do not necessarily reflect the official policy or position of the United States Department of Defense or the U.S. government.

Introduction

David V. Gioe and Margaret W. Smith

For the first two decades of the 21st century, Western strategic thinking, geopolitical orientation, and posturing mainly revolved around counterterrorism and counterinsurgency operations. However, as the Global War on Terror drew to a close, there has been a notable shift in Western, and particularly transatlantic, national security focus toward great power competition. The shift began with the Barak Obama Administration's "pivot to Asia" and the bold moves President Obama took to "shift the center of gravity among the key multilateral organizations in Asia, favoring those that include[d] the United States and leading them to take approaches favored by Washington" (Lieberthal 2011). This shift acknowledged China's increasingly assertive behavior and the realization that U.S. interests should not be narrowly fixated on the Middle East and Central Asia. Recent historical events have underscored the urgency of this strategic reorientation.

The Donald Trump Administration took the Obama pivot further, including in its National Security Strategy the threat of a rising China and an increasingly brazen Russia, arguing that great power competition was back "after being dismissed as a phenomenon of an earlier century" (The White House 2017, p. 27). Western leaders realized that as "China and Russia began to reassert their influence regionally and globally," the role of the United States in international affairs and the U.S.-led global order was being called into question (The White House 2017, p. 27). Especially with the United States and NATO formally withdrawing from Afghanistan in 2021, some have pointed toward waning U.S. influence, greater global multipolarity, and the heretofore unthinkable return of bloody conventional conflict to the European continent. As such, national security experts are revisiting great power competition and focusing on European security, Russia's potential decline, the implications of China's aggressive pursuit of regional hegemony in the Pacific, and, more broadly, Russia and China's goal of reshaping the world order.

This volume is focused on a key dimension of the contemporary great power competition that is often less understood due to its subtle and intangible character: the competition taking place in the cyber domain – broadly conceived – including information and cyber operations. Indeed, democracies across the globe find themselves in an unrelenting cyber and information-based competition with peer and near-peer competitors, but of course, no state is "safe" from the informational

DOI: 10.4324/9781003425304-1

contest. Adversarial powers, particularly China and Russia, recognize that most competition is principally non-kinetic but instead dominates the information environment and cyberspace. This volume articulates the extant Russian and Chinese strategies that elevate cyber and information competition to a central position. Western governments and, in particular, the U.S. government have long conceived of a war-peace duality, but that perspective is giving way to a more nuanced (and useful) perspective on competition. Thus, it seems that the time is ripe to go beyond admiring the problems prevalent in the information space, including but not limited to the cyber domain, and offer a road map for Western powers to compete in and protect the global information environment from malicious actors.

This book's genesis is rooted in the proposition that it is (past) time for the West to push back against authoritarian aggression, but it needs the framework and tools to do so, and we believe that Western democratic states currently lack both strategic and intellectual acumen to compete and win in the informational and cyber domains of great power competition. These deficits may result from an overreliance on conventional military overmatch or overconfidence in a Western-advanced technological offset. But, for whatever reason, we argue that the West needs a roadmap to compete with near-peer powers in information and cyber warfare. With this book, we aim to fill that gap.

To orient the reader to this book, we note that, from an academic perspective, the study of information and cyber operations falls under and cuts across various academic disciplines such as media and communications, sociology, psychology, computer and information sciences, politics, security and defense, intelligence, and many more. We lean into this richness of diverse contributors to reach across disciplinary silos, but we improve upon the academic diversity by noting that most of our chapters are authored by those with practitioner experience as well. Even those contributors currently in academic posts mostly have some practitioner knowledge of the problem set. This very high ratio of practitioners is distinct from other similar books on the subject matter. Indeed, one of the key distinguishing attributes of this volume is the depth, breadth, professional profile, and global reputation of its authors, who cover diverse professional and academic disciplinary grounds.

Specifically, the chapter contributors are lawyers, military officers, defense policy advisors, state department officials, academics, and think tank fellows. In addition to the various fields, the authors offer distinct perspectives beyond American and British, including Western and Eastern European, Baltic, and Latin American. This enables us to provide a range of topics and a balance that may be otherwise hard to find. Beyond the breadth of the authors, the topics are also related but global in orientation. Subsequently, this book's chapters cover the Global South, China, Russia, classic cybersecurity challenges in statecraft, conceptual and societal approaches, to name but a few. Nonetheless, this is done through an explicitly Western lens that is often prescriptive, and all of our authors are writing from a Western perspective and to a democratic audience.

With this volume, we hope to highlight the persistent activities taking place in the information and cyber environments by expanding on the successful article series, *Full-Spectrum: Capabilities and Authorities in Cyber and the Information*

Environment, curated by Barnett Koven and Margaret W. Smith and published by the Modern War Institute (MWI) in partnership with the Army Cyber Institute (ACI) at West Point (The Modern War Institute 2021). This volume builds upon the MWI series to offer additional and expanded chapters written by military officers, government officials, think tank and private sector researchers, and academic experts. We have invited our chapter authors to further develop and integrate their original articles into a robust body of research that explores the strategic employment of the information environment and cyber domain in the context of great power competition as well as possible approaches to respond to – and prevent – malicious activities in cyberspace, ranging from recommendations for policymakers to those aimed at military practitioners and private sector enterprises. The result is a volume of research and experience-based primers for students, mid-career officials, senior leaders in the public and private sectors, and academics alike on key components and conceptualizations of the ongoing information competition from psychological and technical perspectives and an array of response options and recommendations.

The war in Ukraine has loomed large as our authors have been drafting their chapters. Indeed, Russia's (re)invasion of Ukraine, beginning (again) in early 2022, has dominated the headlines and has served as a sort of live demonstration of the integration of information and cyber power with conventional force. Our volume reflects these recent developments. For instance, the chapter by Marina Miron and Rod Thornton considers the strategic logic and efficacy of Russian cyber operations against Ukraine, looking at both the modus operandi as well as the actors involved, and the chapter by David V. Gioe and Tony Manganello considers the democratization of intelligence collection and analysis thanks, in large part, to social media coverage of the war in Ukraine. They argue that the availability of publicly intelligence on the Ukraine war challenges the information "haves" and "have nots," with repercussions for information and operational security in future conflicts. Sandor Fabian and Janis Berzins shift the focus away from Ukraine to offer a perspective from the Baltics that has its gaze firmly fixed on the threats in information and cyber domains emanating from Russia, looking at how these threats evolved from the end of the Cold War.

In placing Russia alongside China for comparative purposes, Chris Marsh and Mark Grzegorzewski address their respective physical, cultural, cognitive, and digital environments to better understand these two near-peer adversaries in contrast and similarity. For both countries – since at least the time of Lenin and Mao – there has only been competition or open conflict. The similarities do not end there. China and Russia place non-kinetic competition at the center of their strategies while also recognizing that competition will largely (though by no means exclusively) play out in third countries. Thus, the chapter aims to inform policymakers to enable them to create strategies to counteract both Russia's and China's influence through concepts like integrated deterrence.

Despite a clear focus on Russia in several chapters, this volume aims to provide insights to help cyber scholars and practitioners in Western democracies conceptualize and contextualize threats, creating a global perspective. The chapter by Tara Heidger and David Higgins explores the African information environment and explains why there is a pressing need for a strategy that considers the African context and information environment, which is currently being penetrated by China and Russia, seeking to dominate the region using mis- and disinformation to discredit Western democracies and even democratic government more broadly. Likewise, the chapter by Fabiana Perera investigates how authoritarian powers seek to achieve their strategic goals across Latin America by exploiting weak governance in the region and harnessing the information and cyber domains.

Moving beyond a regional focus, this volume also considers the challenges of modern warfighting in a contested cyber domain, as covered in the chapters by Christopher Cruden, Otto C. Fiala, and Jim Worrall. In his chapter dedicated to ubiquitous technical surveillance, Cruden warns that contemporary surveillance technologies will challenge military operations in that unintentional trace data will be collected by potential adversaries to learn about the procedures used in highly sensitive operations such as those conducted by Special Operations Forces or to probe for vulnerabilities to exploit in the equipment the U.S. forces use. Fiala and Worrall argue that information and cyber are viewed as a supporting effort for kinetic operations instead of the main effort, focusing the Russian and Chinese operations in the information environment in the Baltics and Taiwan from the perspective of unconventional warfare in the so-called "grey zone," that is, below the threshold of an actual war. They observe that many democratic powers are playing catchup in the competition because, while attention to the cyber domain is increasing, they continue to emphasize a security environment that favors conventional force-on-force engagements during declared conflicts. The chapter by Chaveso Cook analyzes U.S. doctrinal documents related to operations in cyberspace with a particular focus on the United States Cyber Command and military psychological operations. Cook's main argument is that psychological operations can help address some existing Cyber Command's capabilities gaps by leveraging tools like social network analysis (SNA) to better achieve strategic objectives.

This volume also embraces more conceptual or theoretical approaches. For instance, the chapter by Trey Herr, Emma Schroeder and Stewart Scott explains the logic of protraction in cyberspace by looking at irregular violence literature, which helps explain why instability in the modern digital ecosystem can be sustained, while the chapter by Jayson Warren establishes a framework based on quantitative analysis of policy documents to illuminate the concept of competition, drawing attention to the actions of Russia and China that precipitated the U.S. shift away from counterterrorism and toward great power competition. The ultimate aim is to remedy the disjointed policy implementation amid the ongoing competition.

Further, there is a clear normative component to our volume, and this is best exemplified by the chapter authored by Margaret W. Smith and Jim Monken. Shifting the focus from information operations to ransomware attacks, the authors propose a framework for governments to better manage the cybersecurity ecosystem. By looking at the May 2021 Colonial Pipeline hack, the authors demonstrate how the government can cooperate with the private sector to formulate a timely and effective response, thereby deterring cybercriminals.

The final chapters by J.D. Maddox, Casi Gentzel and Adela Levis, and by Michael Poznansky serve to conclude our study of great power competition in the information and cyber domains. Maddox, Gentzel, and Levis offer practical steps to countering disinformation by developing a "whole of society" approach by drawing on examples from the Ukrainian approach to information operations. Their recommendations include rethinking the current regulations and legislation and the role of different public and private actors to adapt both governments and societies to the new reality of disinformation proliferation. Poznansky's chapter concludes this book by looking to the future for ways to respond quickly and credibly to contemporary threats in the cyber domain. Critically, he notes that the challenges covered in this volume will not be solved in a durable way because of the nature of the domain and the actors within it. Indeed, the challenges will endure. Nevertheless, we hope that this volume will help democratic governments and information and cyber practitioners within those states conceptualize the threats, challenges, opportunities, and landscape of information competition heading toward the middle of the 21st century that shows no signs of attenuating conflict in the information environment and cyber domain.

References

Lieberthal, K. G., 2011. The American Pivot to Asia. *Brookings*, 21 December. Available from: www.brookings.edu/articles/the-american-pivot-to-asia/

The Modern War Institute, 2021. *Full-Spectrum: Capabilities and Authorities in Cyber and the Information Environment.* Available from: https://mwi.usma.edu/full-spectrum/

The White House, 2017. *National Security Strategy.* Washington, DC: The White House. Available from: https://trumpwhitehouse.archives.gov/wp-content/uploads/2017/12/NSS-Final-12-18-2017-0905.pdf

1 A Strategic Cyberspace Overview

Russia and China

Mark Grzegorzewski and Christopher Marsh

Introduction

In all U.S. strategy documents, including the latest National Security Strategy (2022) and National Defense Strategy (2022), Russia and China, two revisionist powers, are central to America's strategy to deter and defeat aggression. Analysis of these two competitors and their strategies is therefore a booming business for analysts and practitioners alike. But while Russia's "Gerasimov doctrine," which is not really a doctrine (Galeotti 2019), and China's "unrestricted warfare" make for interesting conjecture, much more significant is perhaps how these two states and their agencies act in cyberspace. Less often discussed and less well understood is that information operations play a central role in both the Russian and the Chinese ways of war, and cyber applications are a central mode by which information is applied as a tool of warfare by both states. China conceives of "informationized warfare" (Fravel 2015), with the space and cyber domains described as becoming the "commanding heights of strategic competition." Despite claims to the contrary, both China and Russia see themselves currently engaged in information warfare against the United States. Prior to Putin's invasion of Ukraine in February 2022, this conflict was in part playing out in the cyber realm. Now that Moscow is waging a full-scale war against its neighbor, cyber has become one of many ways the Kremlin is imposing costs on the West (with energy markets being another important aspect).

The military application of information as an instrument of war – in isolation and in conjunction with other tools – is a central component of both Russia's and China's modern approaches to warfare. As Chief of the Russian General Staff General Valery Gerasimov himself observed, special operations forces leveraging information operations could be effectively employed to "defend and advance [Russia's] national interests beyond" its borders (Kramer 2019). China, for its part, has developed and deployed dedicated information operations units skilled in cyber espionage and cyber-enabled information operations. This chapter serves to highlight some of the differences between how the United States, China, and Russia view cyberspace, and analyze the ways Russia and China are using cyberspace operations to engage the United States asymmetrically.

DOI: 10.4324/9781003425304-2

One way in which both Russia and China view cyberspace operations differently than the United States is through their use of domestic proxies to confront opponents. To the United States, there is a distinct line separating the employment of the state's capabilities from those of private U.S. citizens in cyberspace. Both China and Russia have no qualms about employing commercial companies (Thompson and Lapowsky 2018), "patriotic hackers" (Griffiths 2019), or cybercriminals (Grzegorzewski 2020) on behalf of the state. Among the commercial actors used by Russia and China include the recently deceased Yevgeny Prigozhin-owned Kharkiv News Agency (CSIS 2020) and the Chinese Communist Party (CCP)-associated Sichuan Silence Information organization (BBC 2021) to support their national objectives, and when confronted with evidence of wrongdoing both countries are uncooperative and engage in persistent denial (Ryan et al. 2020; NPR 2021). Such denial is not mere obfuscation, but rather part of their respective deception operations, which both states have long histories employing.

Foremost among the Russian commercial disinformation companies is the troll farm known as the Internet Research Agency (IRA) (Hakala and Melnychuck 2021). Also owned by Prigozhin and based in Saint Petersburg, Russia, the IRA has professionalized information manipulation, creating separate departments to focus on article generation, social media commentary, and multimedia operations. While Russia's intelligence services take the lead in hack-and-dump operations, the IRA serves a related but different mission. The IRA amplifies evocative language and plays on passions to create conflict and disruption in the information space. Chief among the messaging themes used by the IRA is that traditional media and the political system can no longer be trusted. This messaging is executed through what amounts to spamming, meaning if the IRA can get the message out through enough media, the magnitude will overwhelm the information environment. After applying enough pressure to the information sphere, which has already been weakened by attacking its institutional legacy, disinformation in this space just becomes one more piece of information among a plethora of other sources. The goal is that the corrupted narrative becomes as valid a source of information and ultimately trusted alongside traditional institutional knowledge which must compete with foreign information operations. This disruption tears at society's social fabric and causes citizens to lose trust in one another – and trust is a critical component of social movements and collective action, perhaps the only thing that can counter a state's malign information activities.

Of course, both China and Russia leave just enough space between the state and these proxy groups so that they can claim plausible deniability. Regarding so-called "patriotic" hacking, a formal tie may not exist between citizens and the state as the individual actors are willing, autonomous agents acting outside of state control (Davis 2007; Fletcher 2011). However, by not dissuading these irregular actors, it demonstrates that both China and Russia see these actors as furthering their national security interests, and that they likely believe they cannot achieve their aims through traditional diplomacy or even military action.

Another important way in which the United States differs from China and Russia is in how it has organized its military to confront adversaries in cyberspace.

In the United States, the government has divided control over cyberspace. While it has created a military command for the cyberspace domain, U.S. Cyber Command (USCYBERCOM), it also rightfully allows information operations to intersect with each of the other commands. This means there is no particular U.S. military command in charge of information operations as a whole. Rather, all commands share responsibility over this space. Russia and China view cyberspace very differently.

When compared to the United States, Russia, and China claim that traditional notions of sovereignty extend into cyberspace. This view of national and cyberspace sovereignty is undoubtedly influenced by their histories and political cultures. Russia's view of cyberspace sovereignty is shaped by its imperial history. The current Russian Federation (the largest and most powerful of the Soviet successor states) was deeply influenced by the dissolution of two former empires (that of Imperial Russia and the USSR), both of which exacerbated and even created ethnic and nationalistic problems across its territory (Berls 2021). This problem was overlaid with a traditional Kremlin objective of Russian (both Imperial and Soviet) national strategy, to "expand Russia's borders as far from the center as possible until Russian forces reach natural barriers or run into countervailing forces" (Berls 2021). Hence, for more than 800 years, from the time of Muscovy to the Soviet era to the present, Russia's historical experience has conditioned it to view the world in zero-sum terms and to fear the loss of centralized control, including over information and ideas.

To Russia, the information sphere is synonymous with one's territorial sovereignty, and any external attempt to shape the information space is, therefore, a violation of its sovereignty. This view of sovereignty clashes with the U.S. view linked to its national security strategy that sees "the free flow of information online [as something] that enhances international trade and commerce, fosters innovation, and strengthens both national and international security" (The White House 2018). For the United States and Russia, these two claims are irreconcilable. To Russia, physical and information sovereignty is indivisible, while the U.S. views information as yearning to be free and in America's national interest to promote open information spaces, both internally and externally. The Russian concern over the loss of information is so powerful that it claims the U.S. corrupted the minds of Soviet citizens with Western ideas and was behind the "parade of sovereignties" that led to the collapse of the USSR (Kahn 2000).

Russian Cyberspace Operations

For Russia, a core tenet of successful information operations is to be at war with the United States, without Americans even knowing it (and the Kremlin can and does persistently deny it). The Kremlin views cyberspace holistically (Connell and Vogler 2016), to include electronic warfare, psychological operations, and information operations (including information war, or *informatsionnaya voina*).[1] Russia refers to its capabilities as "information weapons" (Ministry of Defense of the Russian Federation 2011), which are essentially anything that broadly affects the human mind, ranging from disinformation to spoofing Global Positioning System (GPS) coordinates to destroying a Programmable Logic Controller (PLC). In Russia, the government has taken a much different view of cyberspace than the U.S. government has, particularly the linking of cyberspace operations to special operations (Smith 2018). The Russian Main Intelligence Directorate (GRU) of the General Staff (the rough equivalent of the U.S. Defense Intelligence Agency, DIA) has primacy in external cyberspace operations, to include espionage, information warfare, and offensive cyberspace operations. This comprehensive approach creates interesting synergies for the Russian military. In addition to the GRU, the Russian Federal Security Service (FSB) has a domestic operations division with an internal security and counterintelligence (CI) mission. The FSB (part of the former KGB) also undertakes external cyberspace operations stemming from its CI responsibilities.

Consequently, its external operations sometimes conflict with the GRU's cyberspace operations due to poor coordination (indeed, poor coordination and C2 plagues Russian special operations in general (Marsh 2017), as seen most clearly in the 2004 Beslan school tragedy (Giduck 2005). As another example, both the GRU and FSB unknowingly targeted the Democratic National Committee (Waterman 2016) at the same time for a hack-and-dump operation. This clearly exemplifies that bureaucratic infighting and friendly fires are not a uniquely American experience. Thankfully for the United States, its competitors are also bureaucratically inept.

The final major Russian agency involved in cyberspace operations is the Foreign Intelligence Service (SVR), which conducts espionage on behalf of the Russian state, and has become quite adept at cyber espionage operations as recently evinced in the SolarWinds hack (Congressional Research Service 2022). The SVR displayed its cyber espionage expertise in this massive attack, in what amounted to a digital supply chain hack. The SVR hacked the IT monitoring and management company, SolarWinds, whose IT performance monitoring system, Orion, provided privileged access to systems. The Orion software was exploited by hacking a third-party vendor and installing malware, which was then pushed out as an update to SolarWinds' 30,000 U.S. clients (18,000 actually installed this update) (Temple-Raston 2021). This malware, pushed out as a genuine update, was not picked up by antivirus software. Once installed, it gave the SVR access into the IT systems of SolarWinds customers and allowed it to move laterally to other networks since the Orion hack provided data on its customer's customers. Dmitri Alperovitch has spoken on the restraint shown by Russia in this hack. Of the 18,000 compromised accounts, only 250 of them have been exploited. To Alperovitch, this allows Russia to hide among the noise by precisely targeted specific accounts while retaining its stealth (CSIS 2021).

As another example of its work, the SVR targeted at least 150 aid groups in May 2021 (BBC 2021). According to Microsoft, the SVR team compromised a genuine USAID e-mail marketing account and in so doing sent out e-mails with a link to organizations in 24 countries. Once the link was clicked, it allowed the SVR to steal data and infect other computers on the network.

Unlike the United States, Russia is not known to have a definitive cyberspace strategy, policy directive, or doctrine (Lilly and Cheravitch 2020). Therefore, what researchers understand about Russian operations and activities in cyberspace is derived from the writings of Russian military scholars and official documents, and even teaching materials at the nation's various military academies.[2] From these sources, it is apparent that the Russian government views cyberspace primarily in terms of "information confrontation" (Helmus et al. 2018) and the technical infrastructure used to control information. To shape and control information, the Russian military takes a "hybrid" approach, integrating "a wide range of overt and covert military, paramilitary, and civilian measures," resulting in a challenging tactical and operational approach (NATO 2014). This includes the use of "special operations forces and non-kinetic political, economic, or informational measures." Therefore, cyber is a central component of the Kremlin's hybrid warfare model, or what Russia refers to as "indirect and asymmetric methods" (Bartles 2016a,b).

Russian information operations are among the best in the world and Russia does not refrain from employing them. Indeed, Russia believes that the tools of information warfare must be brought to bear early and often. To Russia, there is no distinction between information operations, to include those occurring in cyberspace, during times of war or peace (Connell 2017). That said, by employing information operations before the start of a kinetic conflict – as was done in Estonia in 2007 – Russia may be able to achieve its desired strategic aims without having to resort to kinetic military operations. Even where it is still necessary to use kinetic force (e.g., the Russo-Georgian War of 2008), information operations are nevertheless synergistic with the application of military force by, for example, attempting to degrade the resolve of the opposing military force.

Control of information has been critical throughout Russia's authoritarian past, during tsarist times, continuing during the Soviet era, and now under the current authoritarian Putin regime. All these governments came to realize that when they control information, they can also shape the course of events within the country. Conversely, when they lose control of information, such as during the *glasnost'* period, the government cannot control the narrative and may lose legitimacy among the population, resulting in chaos. Therefore, since Russia understands how susceptible it is to losing control over its own information space, it has also come to realize how vulnerable other countries are in this space, especially those countries that have deep societal cleavages. Russia has used its control over domestic media outlets and Russian cyberspace to maintain its control over the population (Hakala 2021). While Russian citizens can purchase virtual private networks (VPN), they are sold by the state, thereby guaranteeing no real privacy while using the VPN. Moreover, because of policy choices made by both the West and Russia, most technology products are indigenously produced in Russia, meaning the Russian state can simply tap into whatever connected devices it likes. In this environment, there is little space for political resistance to Putin or for any political agitation.

While Russia has become very proficient at disinformation campaigns to exploit societal cleavages, it is most adept when it can amplify or augment existing homegrown narratives (Rid 2020). For example, Russia has long sought to exacerbate racial tensions in the United States. Russia was also implicated in a disinformation campaign to discredit the World Anti-Doping Agency (Nimmo et al. n.d.) as a "whole-of-society" approach (the operations include actors beyond the government) to highlight Western moral hypocrisy and undermine international institutions. More recently, Russian information operations headed by either the GRU or the FSB have targeted left-leaning organizations, like Peace Data (Stubbs 2020), and right-leaning platforms, like Parler (Graphika 2020), to exacerbate existing tensions and racial/ethnic cleavages in the United States. While it may seem odd to support both left-leaning and right-leaning organizations, it is part of the Kremlin's plan to simply sow divisive discord among the population as a whole.

Although these actions happen in and through cyberspace, we should not view cyber operations as separate from actions in the physical world. As Martin Libicki (2021) has noted, cyber operations are actually a symptom rather than the disease. Expressed differently, most nation-state acts in and through cyberspace are often in reaction to geopolitical tensions. They are an effort to respond to an action below a level that would trigger armed conflict. This claim finds resonance in the 2021 Russian build-up of military forces along the Ukrainian border. As the Ukrainian government rallied against this build-up of forces, Ukrainian customers, including critical infrastructure, saw a 200% increase in targeting by Russia (Microsoft 2021). Further, the top three countries targeted by Russian cyberspace operations are the United States, the United Kingdom, and Ukraine. Not uncoincidentally, these three countries are also on the list (of eight) countries that Russia considers to be "unfriendly countries" (Microsoft 2021). Cyberspace has become one area in which Russia has capitalized on the asymmetric power of operations and activities within the information space (Jasper 2020). As alluded to previously, the United States and Russia understand the domain and how to employ effects in and through cyberspace very differently. As such, the United States and Russia are fundamentally at odds over any sort of cyberspace "rules of the road" (CSIS 2018). The U.S. push to institutionalize certain behavior in cyberspace (U.S. Department of State 2014; White House 2021) comes from a preference to cement its national interests in cyberspace and keep any future cyberspace rules favorable to U.S. interests. Put another way, a key goal of the United States and other Western liberal democracies is to "maintain a free, stable and open Internet, where fundamental rights and freedoms are ensured" (NATO 2021). This finds expression in the liberal-democratic tradition of openness and transparency and shapes how countries behave and respond in cyberspace (Libicki 2021). For instance, the United States – which is shaped by its own history, experiences, cultures, geography, and norms – has led the post–World War II international order while promulgating the ideology that all countries and peoples yearn to be free. To the United States, this desire to be free is innate and modern borders should not act as an impediment to the spread of freedom. Information should be able to traverse geographic boundaries to allow people to make more informed, and representative, decisions.

This tension between information and control also impacts the United States. In fact, the same belief that information yearns to be free applies within the United

States. In a liberal democracy, a cyber operation is not likely to stay secret for very long. Investigative journalism, public affairs, and requests for information (including through the Freedom of Information Act), all keep the public informed about cyberspace actions that the U.S. government wishes were not public (e.g., Orion breach (New York Times 2021). In contrast, in Russia's authoritarian system of government, control over information is nearly complete. This means that although Russia may suffer from an embarrassing cyberspace operation, it is less likely that this information will ever be admitted publicly. Russia's sovereignty, including in cyberspace, keeps it from acknowledging any adversarial cyberspace action ever took place since any admission would signal that their sovereignty is not complete and that their government is weak. The lack of agreement over cyberspace has particular significance in three areas: espionage, information warfare, and offensive cyberspace operations.

Russian cyberspace espionage is conducted to gather not only intelligence relating to national security, but also economic intelligence. Most recently, Russia was suspected of attempting to hack into pharmaceutical companies in search of COVID-19 research data (Satter 2020). This hacking activity is an unsurprising development to those familiar with Russian strategy. As stated by Gen. Gerasimov (Bartles 2016a,b), Russia must leverage all elements of national power, and this includes cyber espionage, and cyber-enabled economic warfare (Zilberman 2018) to shape the information space and degrade an adversary's capabilities (CRS 2020). One of the more infamous acts of Russian cyber espionage involved a cutout group called the Shadow Brokers (Newman 2018), which likely leveraged the work of the Russian cybersecurity firm Kaspersky to locate National Security Agency (NSA)-developed malware. This malware was possibly found among classified materials that an NSA contractor brought home and operated on his personally owned computer (Shane 2017). The EternalBlue exploits employed by Shadow Brokers subsequently wreaked havoc across the world (Avast 2020).

When it comes to offensive cyberspace operations, or what the U.S. military describes as deny, degrade, disrupt, destroy, and manipulate (D4M) operations, the GRU is the primary actor. The GRU's dominance makes sense given that the SVR and FSB are more focused on espionage. Since the GRU is a military organization, which helps explain the high tolerance for collateral damage, its offensive cyberspace operations are traditionally known for being blunt and reckless, as seen in the NotPetya attack (Greenberg 2018), the Saudi petrochemical attack[*] (Greenberg 2020), or the attacks against Ukraine's power grid (Greenberg 2020). Nonetheless, the GRU has adapted and is now using publicly available penetration tools to hide among the noise and custom malware to avoid detection (Parsons and Raff 2019). The technical sophistication of in-house exploits has allowed Russia to "focus more on advanced intrusion tactics like credential harvesting, supply chain compromises, and infiltrating critical service provider platforms" (Wolff 2021). Despite these advances, Russia still largely prefers to target existing exploits rather than finding new zero days (Wolf 2021), and utilizes a low-tech method of registering domain names similar to popular websites to engage in phishing (Crowdstrike 2019).

Furthermore, in the post–Cold War era, Russia has complained as the North Atlantic Treaty Organization (NATO) has expanded up to its borders, including bringing in the former Soviet republics of Estonia, Latvia, and Lithuania in 2004. Russia views NATO, formerly a collective defense military alliance focused on the threat from the Soviet Union and the Warsaw Pact states of Eastern Europe, as decreasing its security and in fact making conflict more likely. NATO, to Russia, is led by the United States, its main rival, and consequently a genuine threat to its security should a conflict break out (Ziegler 2020). As a result, Russia has employed information operations against NATO to divide it, scanned NATO networks to look for vulnerabilities, and engaged in espionage to determine NATO's intentions (Hakala 2021). The campaign, dubbed Ghostwriter, was an aggressive anti-NATO messaging campaign targeting Lithuania, Latvia, and Poland (FireEye 2019). The campaign spread lies, such as the United States was bringing COVID to Europe during training exercises and that NATO was making the region less secure. The campaign posted to third-party websites using user-submitted content. This content was eventually picked up by legitimate news sources and further spread. The culmination of the campaign was achieved when the Polish War Studies Academy posted a falsified letter on their website calling for the expulsion of U.S. troops from the country (Lyngaas 2020). The information campaign theme, of creating and exacerbating division between European and NATO allies, is not a new theme to the Russians. Rather, they are updating their traditional information campaigns to capitalize on digital interconnectedness by taking social media's scale and amplification qualities to reach a new audience, and then quickly overwhelm them with mis- and disinformation to first establish the narrative. In essence, it is digital judo wherein Russia is taking the West's strength, digital interconnectedness, and reframing that advantage as a weakness and therefore an optimal target.

Russia's cyberspace operations have also hit the United States homeland as in 2018, as the U.S. government accused Russia of running a 2-year intrusion campaign against its power grid (Hlavek 2020). While these cyberspace operations may be D4M operations, others may have been attempts at either mapping networks or preparing the battlefield for usage in wartime. It is simply hard to attribute intent in these cases. It is all about the operator's decision: either use the access for espionage or operations. Cyberspace operations, with their scalability, deniability, and limited resourcing, remain a deliberate tool to challenge the U.S. short of war (NATO 2021). These types of aggressive cyber actions that cause real-world effects will allow Russia to be a formidable force when combined with its traditional strengths in information operations and the insights gleaned through cyber espionage.

Chinese Cyberspace Operations

Much like Russia, China also sees information operations as central to its conception of strategic competition in cyberspace (International Institute for Strategic Studies 2019). Unlike Russia, however, which primarily focuses on an adversary's perception, China focuses more on imposing "adverse physical and technical effects on the opposing system" (Layton 2021). As a point of comparison, from

July 2020 to July 2021, China targeted 13% of U.S. critical infrastructure for potential disruption operations whereas Russia only targeted 2% during that same period (Microsoft 2021). Given this priority, one can conclude that the CCP – and by extension its defender, the People's Liberation Army (PLA) – views information operations via space, cyber, and electronic warfare as the "tip of the spear" in any future conflict to shape the narrative and obtain information superiority, thereby paralyzing a more powerful enemy (Kania and Costello 2018).

China further expanded upon its conceptualization of cyberspace with the creation of the Strategic Support Force (SSF) in 2015, part of China's military rejuvenation campaign with goals such as "build[ing] a world-class army" that can "win modern wars" (People's Liberation Army of China 2013). Some analysts view the SSF as an enhanced Chinese counterpart to USCYBERCOM. The SSF not only focuses on the traditional D4M operations of USCYBERCOM, but has also added space, electronic, and psychological warfare to their repertoire. Housing these different but complementary cyber-enabled capabilities within the same command is expected to create synergies that these capabilities cannot achieve on their own (Ng 2020). Moreover, having a suite of functions under the same command during peacetime will give the CCP and PLA the ability to seamlessly transition to an integrated campaign during wartime (Kania et al. 2018). Unlike the SSF – which focuses more on wartime capabilities – the Ministry of State Security (MSS), which has grown significantly under Xi Jinping, focuses primarily on economic espionage and fuels China's "private sector" innovation (The Economist 2021).

While China has not openly published a cyberspace strategy, scholars and practitioners are in widespread agreement as to the CCP's aims. It seeks to control the flow of information to and within China to ensure domestic stability (Chang 2014) (and halt the efforts of "splittists" [separatists] who seek the disintegration of the PRC), and preserve economic growth through commercial espionage (Segal 2018). By controlling dissent and driving economic growth, the CCP seeks to ensure that it maintains power.

China views the information space as just one component within cyberspace. For the CCP, to remain in power, it must assert complete dominance over its information space (Beecroft 2021). This creates opportunities for cyberspace investments by CCP-friendly Internet companies and allows the CCP to retain complete control over anything it finds objectionable in its information space (Associated Press 2021). This digital surveillance architecture scans the Chinese information space for any promotion of the "Three Evils": terrorism, extremism, and separatism (Richards 2021).

China asserts that just as every state is the sovereign within its own borders, each should likewise be the sovereign within its own cyberspace (Dou 2016). Cyber sovereignty challenges the U.S. view that information should be allowed to flow freely across borders (Chang 2014). China considers the control of information internally to be as vital as "controlling the maritime domain in the eighteenth century or controlling the air domain in the twentieth century" (Kolton 2017). Therefore, to maintain harmony within China and produce disruptive effects beyond its borders, China has increasingly improved its information operations throughout recent years.

Regarding sustaining economic growth, China's operations against economic targets and the commercial sector are viewed by former USCYBERCOM commander General Keith Alexander as "the greatest transfer of wealth in history" (Alexander 2012). Through the complete "informationalization" of their economy, which completely intertwines the CCP, China's economy, and Chinese society, the Chinese economy continues to grow (Beecroft 2021). Informationalization extends from the CCP's control over domestic social media spaces to the new Digital Currency Electronic Payment to social credit scores, which all serve the dual purpose of controlling information internally and promoting Chinese economic interests (Harvard 2021).

While the CCP continues to claim that cyber economic espionage is not the work of the government but rather criminal elements within China (Chang 2014), the cybersecurity group FireEye (FireEye 2021) has been able to identify with a high degree of certainty that there are at least ten advanced persistent threats (APTs) operated by the CCP, nine of which focus on industrial espionage. Moreover, China has left just enough plausible deniability and Chinese hackers to feign ignorance. Hackers emanating from China targeting intellectual property may in some cases work for front companies, contractors, or maybe security researchers (Perlroth 2021). It is unknown whether Chinese hackers are receiving guidance from the CCP, or whether they are moonlighting while the state turns a blind eye to their activities.

In addition to these government-supported APTs, China also has a very large patriotic hacker community that it can mobilize when needed (Siboni 2012). Due to the extensive cyberspace dragnet that the CCP has put in place, the government is aware of the activities of these hackers and can stop them when it so desires (Chang 2014). However, the CCP has also employed this hacker network as an extension of the state while at the same time retaining plausible deniability since these patriotic hackers are not formally part of the state apparatus (Applegate 2011). Due to the strong sense of Chinese nationalism, coupled with historical grievances over imperialism, many Chinese patriotic hackers feel a strong sense to "defend" their homeland and push back against what they perceive to be modern Western imperialism (Richards 2021). As a consequence, patriotic hackers are viewed as national heroes within their country for "restor[ing] China's honor, glory, and integrity" (Kassner 2013).

The CCP's firm hold on power is a function of both its ability to maintain economic growth and its control over the flow of information in China. This explains why both economic cyberspace espionage and information operations have seen increased investment by the Chinese military in recent years (U.S. Department of Defense 2018). Moreover, the CCP has leveraged China's commercial sector to support the state's interests, as seen by the "Made in China 2025" campaign that encourages Chinese companies to create dual-use technologies, which can also be employed by the military (U.S. Department of Defense 2018). This coupling of the state and private sector is seen in China's Cyber Security Law and National

Intelligence Law, the latter of which necessitates that "any organization and citizen shall, in accordance with the law, support, provide assistance, and cooperate in national intelligence work, and guard the secrecy of any national intelligence work that they are aware of" (Hoffman et al. 2018).

China has also begun setting international standards for digital technology (Pop et al. 2021). Filling the international bodies that the U.S. has long since abandoned, the CCP has found value in shaping international institutions in ways that favor China's national interests. Standard setting serves the CCP's interest by both allowing it to regulate the framework that technology companies will use both to manufacture and control information technology. This technology is both employed by China and exported around the world, often as part of the Belt and Road Initiative (BRI), are part of China's digital authoritarian model to control information and push for digital sovereignty (Sherman 2021; Attrill et al. 2021).

The pervasiveness of the CCP attempt to control information, both foreign and domestic, can be found in its recent deployment of social media surveillance technology (Cadell 2021). The invasive and pervasive technology deployed by the CCP mines social media accounts to create a database of foreigners that comment on hot-button issues such as Hong Kong, Taiwan, and the Uyghurs. Ultimately, this surveillance technology, coupled with the estimated $7 billion the CCP spends annually on Internet censorship (Inkster 2022), allows the CCP to both monitor the external and internal information environment and shape how its citizens – and the world – view China.

The Chinese state does not employ externally focused information operations nearly as effectively as their Russian counterparts. However, the Chinese state is constantly learning, and has likely gleaned a lot from Russia's successful 2016 influence operations in the U.S. political space (Brandt and Schafer 2020). Traditionally, the Chinese state focused most of its information operations internally to maintain domestic and regime stability. This made tremendous sense in the wake of both the 1989 Tiananmen Square demonstrations and the 1991 collapse of the Soviet Union (which feared a similar fate; Marsh 2006). However, this slowed under Hu Jintao but then exploded under the rise to power of Xi Jinping, now in his unprecedented third term. Under Xi, China has shifted some of its information operations from being used primarily for domestic control to influencing the external environment (Diresta et al. 2020). This is most evident in China's projection of a newfound muscular international image (dubbed "wolf warrior diplomacy" after the record-setting Chinese film franchise) (Brandt and Taussig 2020), and in its defensiveness over COVID-19. China seeks to influence the environment post–COVID by demonstrating that its top-down authoritarian methods are superior to the West's deliberative, and messy, approach to governing (Manantan 2021). One asymmetric tool of the CCP to covertly push its preferred narrative are the TikTok (ByteDance) and WeChat (Tencent) platforms (Ryan et al. 2020). Both companies have internal CCP committees "to ensure that the party's political goals are pursued alongside the companies' commercial goals" (Ryan et al. 2020).

Moreover, as China attempts to increase its security in Asia (Wuthnow 2017), it has started targeting countries in the Mekong Region (Cambodia, Laos, Myanmar,

Thailand, and Vietnam) to shape their views of China and China's foreign policy (Holz and Loomis 2020). The CCP social media narratives have focused on "emphasiz[ing] China's historical, cultural, political, and geographic ties" as well as the economic benefits of partnering with China (Holz et al. 2020). Outside of its region, China has also invested in telling its "story" well by targeting more foreign social media platforms with its own media, scholars, and officials (Bachman et al. 2020). Chinese large-scale, external influence operations still need further refinement, especially in employing Deepfake technologies (Nimmo et al. 2019). Recently, CCP external influence operations have had some sporadic successes, due to increased persona development, better resonating narratives, and sheer volume (Nimmo et al. 2021). That said, as China continues to test and develop its influence campaigns through cyberspace, we can expect the content of their messaging to improve and their impact on foreign audiences to be greater.

Looking Ahead

Both Russia and China view information and cyberspace operations differently than the United States, and they are designing their operations and cyber infrastructure to engage the United States asymmetrically. In Russia, such asymmetric and nonlinear means are most clearly articulated by Gen. Valery Gerasimov, who has stated that "asymmetric warfare" can produce effects "as significant as conventional operations" (Gerasimov 2016). In China, it is the 2013 *Science of Military Strategy* that most clearly encapsulates such ideas. Citing Sun Tzu's emphasis on *fei duicheng* ("asymmetric means"), such measures are to be employed against adversaries with an emphasis on China developing its "special asymmetric, contactless, and nonlinear warfare style" (People's Liberation Army of China 2013). It is clear that the Sino-Russian "no-limits" strategic partnership has extended into the realm of conceptual development at the operational and tactical levels (Burke et al. 2020).

The United States needs to more fully appreciate the contemporary operational and strategic environments to take more meaningful action, whether offensive or defensive, in cyberspace when dealing with Russia and China. As it stands, the U.S. government is taking action against its adversaries without fully appreciating their view of the operational environment. This helps explain why U.S. attempts to influence the environment do not always work. The U.S. government often has little to no understanding of how the adversary views itself and its interests, and therefore the target of an effect is less impactful since the U.S. does not even know its adversary's starting position. We started with the critical premise that Russia and China are currently at war with the United States, albeit using different and unconventional means. Until the U.S. national security community accepts that underlying fact, the long list of recommended cyberspace actions against its adversaries are worthless. The military application of information as an instrument of war – in isolation and in conjunction with other tools – is a central component of these states' modern approaches to warfare, both today and into the foreseeable future. The United States must stop viewing these information tools in isolation. They are capabilities that these adversaries are using to signal and register their views to the

United States regarding the current geopolitical environment. Recognition of this reality must undergird America's cyberspace and information warfare policies and doctrine. Cyberspace is not disconnected from the physical space, and actions of U.S. adversaries do not occur in a vacuum. It is time the United States stops being so enamored with cyberspace effects and starts asking why adversaries are acting in the environment and why they are choosing cyberspace-based effects. Only then can the U.S. combine physical and cyber effects to effectively combat and deter its adversaries.

In closing, the 2022 National Defense Strategy specifically calls out "integrated deterrence" as the framework for working across warfighting domains, theaters, and the spectrum of conflict. The Department of Defense (DoD) plays an integral role in integrated deterrence, and yet the DoD could be even more impactful, specifically in and through cyberspace, with a deeper understanding of deterrence. In many U.S. government organizations, there is a profound lack of understanding regarding the conceptual underpinnings that inform the deterrence framework. Among several conceptual factors for integrated deterrence to be effective, the U.S. must understand what motivates its competitors and how its competitors view their world. Otherwise, effects in and through cyberspace may not have the intended effect and may well send unintended signals thereby deepening misperception, undermining integrated deterrence, and possibly escalating crises. Focusing on America's primary competitors, Russia and China, this chapter addressed the physical, cultural, cognitive, and digital environments to better understand the Putin regime and the CCP. Ideally, this focus will better inform policymakers on how Russia and China view themselves, outsiders, and competition and enable the United States to tailor specific responses in and through cyberspace for integrated deterrence.

Notes

1 There is no word in Russian for "warfare," only "war." Hence, there is arguably no cognitive distinction between what is considered a method of engaging in war (warfare) and outright war itself.
2 See, for example, the work of Thomas (2006).

References

Alexander, K., 2012. *Gen. Alexander: Greatest Transfer of Wealth in History*. American Enterprise Institute. Available from: www.youtube.com/watch?v=JOFk44yy6IQ

Applegate, S., 2011. *Cybermilitias and Political Hackers: Use of Irregular Forces in Cyberwarfare*. IEEE. Available from: https://ieeexplore.ieee.org/document/5765925

Associated Press, 2021. *China's Communist Party Exerting Tighter Control Over the Country's Internet Giants*. Available from: www.marketwatch.com/story/chinas-communist-party-exerting-tighter-control-over-the-countrys-internet-giants-01633320070

Attrill, N. and Fritz, A., 2021. *China's Cyber Vision How the Cyberspace Administration of China is Building a New Consensus on Global Internet Governance*. Australia Strategic Policy Institute. Available from: www.aspi.org.au/report/chinas-cyber-vision-how-cyberspace-administration-china-building-new-consensus-global

Avast, 2020. *C-eternalblue*. Available from: www.avast.com/c-eternalblue

Bachman, E. and Bellacqua, J., 2020. *Black and White and Red All Over: Chinas Improving Foreign-Directed Media*. Arlington, VA: CNA. Available from: www.cna.org/reports/2020/08/chinas-improving-foreign-directed-media

Bartles, C. K., 2016a. *Getting Gerasimov Right*. Army University Press. Available from: www.armyupress.army.mil/Portals/7/military-review/Archives/English/Military Review_20160228_art009.pdf

Bartles, C. K., 2016b. Russia's Indirect and Asymmetric Methods as a Response to the New Western Way of War. *Special Operations Journal*, 2(1), 1–11.

BBC, 2021. *Facebook Uncovers Chinese Network Behind Fake Expert*. BBC. Available from: www.bbc.com/news/world-asia-china-59456548

Beecroft, N., 2021. *The West Should Not Be Complacent About China's Cyber Capabilities*. Carnegie Endowment for International Peace. Available from: https://carnegieendowment.org/2021/07/06/west-should-not-be-complacent-about-china-s-cyber-capabilities-pub-84884

Berls, R., 2021. *Survival of the Russian State: Protecting It from Foreign and Domestic Threats, Part I*. Available from: www.nti.org/analysis/articles/survival-russian-state-protecting-it-foreign-and-domestic-threats-part-i/

Brandt, J. and Schafer, B., 2020. *How China's 'Wolf Warrior' Diplomats Use and Abuse Twitter*. Brookings. Available from: www.brookings.edu/techstream/how-chinas-wolf-warrior-diplomats-use-and-abuse-twitter/

Brandt, J. and Taussig, T., 2020. *The Kremlin's Disinformation Playbook Goes to Beijing*. Brookings Institution. Available from: www.brookings.edu/blog/order-from-chaos/2020/05/19/the-kremlins-disinformation-playbook-goes-to-beijing/

Burke, E., Gunness, K. Cooper III, C., and Cozad, M., 2020. *People's Liberation Army Operational Concepts*. Santa Monica, CA: RAND.

Cadell, C., 2021. China Harvests Masses of Data on Western Targets, Documents Show. *Washington Post*, 31 December. Available from: www.washingtonpost.com/national-security/china-harvests-masses-of-data-on-western-targets-documents-show/2021/12/31/3981ce9c-538e-11ec-8927-c396fa861a71_story.html

Chang, A., 2014. *Warring State: China's Cybersecurity Strategy*. Center for a New American Security. Available from: www.cnas.org/publications/reports/warring-state-chinas-cybersecurity-strategy

Congressional Research Service (CRS), 2020. *Russian Armed Forces: Military Doctrine and Strategy*. Available from: https://fas.org/sgp/crs/row/IF11625.pdf

Congressional Research Service (CRS), 2022. *Russian Cyber Units*. Washington, DC: CRS. Available from https://crsreports.congress.gov/product/pdf/IF/IF11718

Connell, M., 2017. *Russia's Approach to Cyber Warfare*. Center for Naval Analysis. Available from: www.cna.org/cna_files/pdf/DOP-2016-U-014231-1Rev.pdf

Connell, M. and Vogler, S., 2016. *Russia's Approach to Cyber Warfare*. New York: Center for Naval Analysis. Available from: www.cna.org/cna_files/pdf/DOP-2016-U-014231-1Rev.pdf

Crowdstrike. 2019. *Who is FANCY BEAR (APT28)?* Available from: www.crowdstrike.com/blog/who-is-fancy-bear/

CSIS, 2018. *Responding to Russia: Deterring Russian Cyber and Grey Zone Activities*. Washington, DC: CSIS. Available from: www.csis.org/analysis/responding-russia-deterring-russian-cyber-and-grey-zone-activities

CSIS, 2020. *Moscow's Mercenary Wars*. Washington, DC: CSIS. Available from: https://russianpmcs.csis.org/

CSIS, 2021. *Evolution of Russian Cyber Tactics and Operations.* Washington, DC: CSIS. Available from: www.csis.org/events/evolution-russian-cyber-tactics-and-operations

Davis, J., 2007. *Hackers Take Down the Most Wired Country in Europe.* San Francisco: Wired. Available from: www.wired.com/2007/08/ff-estonia/

Diresta, R., Miller, C., Molter, V., Pomfret, J., and Tiffert, G., 2020. *Telling China's Story: The Chinese Communist Party's Campaign to Shape Global Narratives.* Hoover Institution. Available from: https://fsi-live.s3.us-west-1.amazonaws.com/s3fs-public/sio-china_story_white_paper-final.pdf

Dou, E., 2016. China's Cyber Strategy Stresses Securing Infrastructure. *The Wall Street Journal,* 27 December. Available from: www.wsj.com/articles/chinas-cyber-strategy-stresses-securing-infrastructure-1482842486

FireEye, 2019. *'Ghostwriter' Influence Campaign.* Available from: www.fireeye.com/content/dam/fireeye-www/blog/pdfs/Ghostwriter-Influence-Campaign.pdf

FireEye, 2021. *M-Trends.* Available from: https://content.fireeye.com/m-trends/rpt-m-trends-2020

Fletcher, O., 2011. Patriotic Chinese Hacking Group Reboots. *The Wall Street Journal,* 5 October. Available from: www.wsj.com/articles/BL-CJB-14435

Fravel, M., 2015. *China's New Military Strategy: "Winning Informationized Local Wars."* Washington, DC: Jamestown Foundation. Available from: https://jamestown.org/program/chinas-new-military-strategy-winning-informationized-local-wars/

Galeotti, M., 2019. *I'm Sorry for Creating the 'Gerasimov Doctrine.'* Washington, DC: Foreign Policy. Available from: https://foreignpolicy.com/2018/03/05/im-sorry-for-creating-the-gerasimov-doctrine/

Gerasimov. V., 2016. Po Opytu Sirii [On the Syrian Experience]. *Voennoe-Promyshlennyi Kur'er,* 9(624), 9 March. Available from: https://vpk.name/news/150974_po_opytu_sirii.html

Giduck, J., 2005. *Terror at Beslan: A Russian Tragedy with Lessons for America's Schools.* Olympia, WA: Archangel Group Inc.

Graphika, 2020. *Step into My Parler.* Available from: https://public-assets.graphika.com/reports/graphika_report_step_into_my_parler.pdf

Greenberg, A., 2018. The Untold Story of NotPetya, the Most Devastating Cyberattack in History. *Wired,* 22 August. Available from: www.wired.com/story/notpetya-cyberattack-ukraine-russia-code-crashed-the-world/

Greenberg, A., 2020. The US Sanctions Russians for Potentially 'Fatal' Triton Malware. *Wired,* 23 October. Available from www.wired.com/story/russia-sanctions-triton-malware/

Griffiths, J., 2019. When Chinese Hackers Declared War on the Rest of Us. *MIT Technology Review,* 10 January. Available from www.technologyreview.com/2019/01/10/103560/when-chinese-hackers-declared-war-on-the-rest-of-us/

Grzegorzewski, M., 2020. Russian Cyber Operations: The Relationship between the State and Cybercriminals. *In: Cyber Terrorism and Extremism as Threat to Critical Infrastructure Protection.* Publishing Houses Ministry of Defence Republic of Slovenia, Joint Special Operations University from Tampa, USA and Institute for Corporative Security Studies, Ljubljana, Slovenia. Available from: https://dk.mors.si/Dokument.php?id=1801&lang=slv

Hakala, J. and Melnychuck, J., 2021. *Russia's Strategy in Cyberspace.* NATO Strategic Communications Centre of Excellence. Available from: https://stratcomcoe.org/publications/russias-strategy-in-cyberspace/210

Harvard, 2021. *Made in China 2025 Explained.* Available from: https://perma.cc/W6L5-RVPN

Helmus, T.C., Bodine-Baron, E., Radin, A., Magnuson, M., Mendelsohn, J., Marcellino, W., Bega, A. and Winkelman, Z., 2018. *Russian Social Media Influence: Understanding Russian Propaganda in Eastern Europe.* Santa Monica, CA: RAND Corporation.

Hlavek, A., 2020. The Russian Threat, In Brief. *Ironnet.* Available from: https://f.hubspotus ercontent20.net/hubfs/6306975/PDFs/Russia%20Cyber%20Threat%20Report%20Octo ber%202020.pdf

Holz, H. and Loomis, R., 2020. *China's Efforts to Shape the Information Environment in the Mekong Region.* Arlington, VA: Center for Naval Analysis (CAN). Available from: www. cna.org/reports/2021/10/IIM-2020-U-027917-Final.pdf

Inkster, N., 2022. Power Versus Pragmatism. *The Cyber Defense Review,* 7(1), 41–50.

International Institute for Strategic Studies, 2019. *China's Cyber Power in a New Era In Asia Pacific Regional Security Assessment 2019.* Available from: www.iiss.org/publicati ons/strategic-dossiers/asiapacific-regional-security-assessment-2019/rsa19-07-chapter-5

Jasper, S., 2020. *Russian Cyber Operations: Coding the Boundaries of Conflict.* Washington, DC: Georgetown University Press.

Kahn, J., 2000. The Parade of Sovereignties: Establishing the Vocabulary of the New Russian Federalism. *Post-Soviet Affairs,* 16 (1): 58–89.

Kania, E.B. and Costello, J.K., 2018. The Strategic Support Force and the Future of Chinese Information Operations. *The Cyber Defense Review,* 3(1), 105–122.

Kassner, M. 2013. *Understanding What Motivates Chinese Hackers.* Available at www.techr epublic.com/blog/it-security/understanding-what-motivates-chinese-hackers/

Kolton, M., 2017. Interpreting China's Pursuit of Cyber Sovereignty and its Views on Cyber Deterrence. *The Cyber Defense Review,* 2(1), 119–154.

Kramer, A. 2019. Russian General Pitches 'Information' Operations as a Form of War. *New York Times.* Available at www.nytimes.com/2019/03/02/world/europe/russia-hybrid-war-gerasimov.html

Layton, P., 2021. Fighting Artificial Intelligence Battles: Operational Concepts for Future AI-Enabled Wars. *Network,* 4(20), 1–100.

Libicki, M.C., 2021. *Cyberspace in Peace and War.* Annapolis, MD: Naval Institute Press.

Lilly, B. and Cheravitch, J., 2020. *The Past, Present, and Future of Russia's Cyber Strategy and Forces.* Santa Monica, CA: RAND Corporation. Available from: www.rand.org/pubs/external_publications/EP68319.html

Lucas, R., 2020. DOJ Charges Chinese Nationals with Hacking More Than 100 Companies. *NPR,* 16 September. Available from: www.npr.org/2020/09/16/913618435/doj-charges-chinese-nationals-in-allegedly-hacking-of-more-than-100-companies

Lyngaas, S., 2020. Poland Implicates Russia in Cyberattack, info OP Aimed at Undercutting US Relations. *Cyberscoop,* 24 April. Available from: https://cyberscoop.com/poland-cyberattack-russia-us-military/

Manantan, M.B., 2021. Unleash the Dragon. *The Cyber Defense Review,* 6(2), 71–90.

Marsh, C., 2006. *Unparalleled Reforms: China's Rise, Russia's Fall, and the Interdependence of Transition.* Lanham, MD: Lexington.

Marsh, C., 2017. *Developments in Russian Special Operations: Russia's Spetsnaz, SOF, and Special Operations Forces Command.* Kingston, ON: Canadian Special Operations Forces Command Monograph.

Microsoft, 2021. *Microsoft Digital Defense Report.* Available from: https://query.prod.cms. rt.microsoft.com/cms/api/am/binary/RWMFIi?id=101738

Ministry of Defense of the Russian Federation, 2011. *Conceptual Views on the Activities of the Russian Federation Armed Forces in the Information Space.* Mil.ru. Available from: https://eng.mil.ru/en/science/publications/more.htm?id=10845074@cmsArticle

NATO, 2014. *Wales Summit Declaration.* 5 September. Available from: www.nato.int/cps/en/natohq/official_texts_112964.htm

NATO Strategic Communications Centre of Excellence, 2021. *Russia's Strategy in Cyberspace.* Available from: https://stratcomcoe.org/publications/russias-strategy-in-cyberspace/210

Newman, L., 2018. The Leaked NSA Spy Tool that Hacked the World. *Wired,* 7 March. Available from: www.wired.com/story/eternalblue-leaked-nsa-spy-tool-hacked-world/

New York Times, 2021. Russian Hackers Broke into Federal Agencies, U.S. Officials Suspect. *New York Times,* 13 December. Available from: www.nytimes.com/2020/12/13/us/politics/russian-hackers-us-government-treasury-commerce.html

Ng, J., 2020. China Broadens Cyber Options. *Asian Military Review,* 15 January. Available from: https://asianmilitaryreview.com/2020/01/china-broadens-cyber-options/

Nimmo, B., Eib, C., and Tamora, L., 2019. *Cross Platform Spam-Network Targeted Hong Kong Protests.* New York: Graphika. Available from: https://public-assets.graphika.com/reports/graphika_report_spamouflage.pdf

Nimmo, B., Francois, C., Eib, C., Ronzaud, L., Ferreira, R., Hernon, C., and Kostelancik, T., n.d. *Secondary Inkfektion.* New York: Graphika. Available from: https://secondaryinfektion.org/download

Nimmo, B., Hubert, I. and Cheng, Y., 2021. *Spamouflage Breakout: Chinese Spam Network Finally Starts to Gain Some Traction.* New York: Graphika. Available from: www.graphika.com/reports/spamouflage-breakout/

NPR, 2021. Russian Cyberattacks Present Serious Threat to U.S. *NPR Morning Edition,* 9 July. Available from: www.npr.org/2021/07/09/1014512241/russian-cyber-attacks-present-serious-threat-to-u-s

Parsons, E. and Raff, M., 2019. *Understanding the Cyber Threat from Russia.* F-Secure Cyber Security. Available from: www.f-secure.com/en/consulting/our-thinking/understanding-the-cyber-threat-from-russia

People's Liberation Army of China, 2013. *The Science of Military Strategy.* Beijing: Military Science Publishing House.

Perlroth, N., 2021. How China Transformed into a Prime Cyber Threat to the U.S. *The New York Times,* 19 July. Available from: www.nytimes.com/2021/07/19/technology/china-hacking-us.html

Pop, V., Hu, S. and Michaels, D., 2021. From Lightbulbs to 5G, China Battles West for Control of Vital Technology Standards. *The Wall Street Journal,* 8 February. Available from: www.wsj.com/articles/from-lightbulbs-to-5g-china-battles-west-for-control-of-vital-technology-standards-11612722698

President, 2022. *National Security Strategy.* Washington, DC: White House. Available from: www.whitehouse.gov/wp-content/uploads/2022/10/Biden-Harris-Administrations-National-Security-Strategy-10.2022.pdf

Richards, A., 2021. Evolution of China's Cyber Threat. *Small Wars Journal,* 23 September. Available from: https://smallwarsjournal.com/jrnl/art/evolution-chinas-cyber-threat

Rid, T., 2020. *Active Measures: The Secret History of Disinformation and Political Warfare.* New York: Farrar, Straus and Giroux.

Ryan, F., Fritz, A., and Impiombato, D., 2020. TikTok and WeChat. *Australian Strategic Policy Institute,* 8 September. Available from: www.aspi.org.au/report/tiktok-wechat

Satter, R., 2020. North Korean, Russian Hackers Target COVID-19 Researchers: Microsoft. *Reuters*, 13 November. Available from: www.reuters.com/article/us-microsoft-coronavi rus-cyber/north-korean-russian-hackers-target-covid-19-researchers-microsoft-idUSKB N27T1WF

Segal, A., 2018. When China Rules the Web: Technology in Service of the State Foreign Affairs. *Foreign Affairs*, 13 August. Available from: www.foreignaffairs.com/articles/ china/2018-08-13/when-china-rules-web

Shane, S., 2017. Malware Case Is Major Blow for the N.S.A. *The New York Times*, 16 May. Available from: www.nytimes.com/2017/05/16/us/nsa-malware-case-shadow-brok ers.html

Sherman, J., 2021. Digital Authoritarianism and Implications for US National Security. *The Cyber Defense Review*, 6(1), 107–118.

Siboni, G.Y.R., 2012. What Lies Behind Chinese Cyber Warfare. *Military and Strategic Affairs*, 4(2), 49–64.

Smith, B., 2018. *Russian Intelligence Services and Special Forces*. London: The House of Commons Library. Available from: https://fas.org/irp/world/russia/CBP-8430.pdf

Stubbs, J., 2020. Duped by Russia, Freelancers Ensnared in Disinformation Campaign by Promise of Easy Money. *Reuters*, 2 September. Available from: www.reuters.com/article/ us-usa-election-facebook-russia/duped-by-russia-freelancers-ensnared-in-disinformat ion-campaign-by-promise-of-easy-money-idUSKBN25T35E

Temple-Raston, D., 2021. A 'Worst Nightmare' Cyberattack: The Untold Story of The SolarWinds Hack. *NPR*, 16 April. Available from: www.npr.org/2021/04/16/985439655/ a-worst-nightmare-cyberattack-the-untold-story-of-the-solarwinds-hack

The Economist, 2021. State Sponsored Hacking: The Spectral Game. *The Economist*, 11 November. Available from: www.economist.com/china/2021/11/11/china-still-steals-com mercial-secrets-for-its-own-firms-profit

Thomas, T., 2006. *Cyber Silhouettes: Shadows Over Information Operations*. Fort Leavenworth, KS: Foreign Military Studies Office.

Thompson, K. and Lapowsky, J., 2018. How Russian Trolls Used Meme Warfare to Divide America. *Wired*, 17 December. Available from: www.wired.com/story/russia-ira-propaga nda-senate-report/

U.S. Department of Defense, 2018. *Annual Report to Congress: Military and Strategic Developments Involving the People's Republic of China 2018*. Available from: https://media. defense.gov/2018/Aug/16/2001955282/-1/-1/1/2018-CHINA-MILITARY-POWER-REP ORT.PDF

U.S. Department of Defense, 2022. *National Defense Strategy of the United States of America*. Washington, D.C.: Department of Defense. Available from: https://media.defe nse.gov/2022/Oct/27/2003103845/-1/-1/1/2022-NATIONAL-DEFENSE-STRATEGY-NPR-MDR.PDF

U.S. Department of State, 2014. *Report on A Framework for International Cyber Stability*. Available from: https://2009-2017.state.gov/documents/organization/229235.pdf

Waterman, S., 2016. U.S. Sanctions Russian Spy Agencies, Officials for Hacking. *Cyberscoop*, 29 December. Available from: www.cyberscoop.com/russia-obama-sancti ons-dnc-hacks-gru-fsb/

White House, 2018. *National Cyber Strategy*. Available from: https://trumpwhitehouse. archives.gov/wp-content/uploads/2018/09/National-Cyber-Strategy.pdf

White House, 2021. *The United States, Joined by Allies and Partners, Attributes Malicious Cyber Activity and Irresponsible State Behavior to the People's Republic of China.*

Available from: www.whitehouse.gov/briefing-room/statements-releases/2021/07/19/the-united-states-joined-by-allies-and-partners-attributes-malicious-cyber-activity-and-irresponsible-state-behavior-to-the-peoples-republic-of-china/

Wolff, J., 2021. Understanding Russia's Cyber Strategy. *Foreign Policy Research Institute*, 6 July. Available from: www.fpri.org/article/2021/07/understanding-russias-cyber-strategy/

Wuthnow, J., 2017. *Chinese Perspectives on the Belt and Road Initiative: Strategic Rationales, Risks, and Implications*. Washington, DC: National Defense University Press.

Ziegler, C.E., 2020. A Crisis of Diverging Perspectives: US-Russian Relations and the Security Dilemma. *Texas National Security Review*, 4(1), 12–33.

Zilberman, B., 2018. Don't Underestimate Economic Side of Russia's Cyber Warfare. *The Cipher Brief*, 25 June. Available from: www.thecipherbrief.com/column_article/dont-underestimate-economic-side-russias-cyber-warfare

2 On Competition

A Continuation of Policy by Misunderstood Means

Jayson Warren

Introduction

American political scholars often characterize the policymaking process in terms of incrementalism (Birkland 2020, p. 298)[1] – policy via small changes over time – and generally attribute this governmental tendency to the Founders' diffusion of power across national, state, and local levels. When incrementalism is disrupted, it is described as punctuated equilibrium (Baumgartner and Jones 2009, pp. 18–19) – a term borrowed from biological evolution to describe instances where stable systems are decisively interrupted by significant alterations. Punctuated equilibrium at the nexus of national security and foreign policy often coincides with a focusing event (Kingdon 2011, pp. 94–100), a shock to the system that not only demands expedient large-scale change to the status quo but by extension also serves as a unifying catalyst for bipartisan cooperation and societal support (see Table 2.1).

However, the most recent instance of punctuated equilibrium within foreign policy (i.e., the Trump administration's pivot to great power competition) emerged in such a way that the focusing event was enigmatic, fostering partisan conflict and inconsistent societal support, which in turn disrupted the policy's implementation. Although the Obama administration's 2015 National Security Strategy (NSS) did assert that "China's rise, and Russia's aggression [both] significantly impact the future of major power relations" (U.S. Executive Office of the President 2015, p. 4), the tones struck by the Trump NSS were considered a significant point of departure from precedent – particularly it is a core assertion: "After being dismissed as a phenomenon of an earlier century, great power competition returned" (the emphasis here is the past-tense verb *returned*, which connotes someone other than the United States brought it onto the world stage; U.S. Executive Office of the President 2017, p. 27). A few months later, Secretary of Defense James Mattis doubled down on the President's position in his National Defense Strategy (NDS): "Inter-state strategic competition, not terrorism, is now the primary concern in U.S. national security" (U.S. Department of Defense 2018, p. 1).

Although it goes without saying competition as policy had numerous supporters across government, industry, and academia, to many observers the policy emerged without context or even a clear need, thus leaving policy commentators to avoid

DOI: 10.4324/9781003425304-3

Table 2.1 Focusing Events and Corresponding Punctuated Equilibrium

Focusing event	Punctuated equilibrium in policy environment
Japan attacks Pearl Harbor	U.S. reversed its policy of isolationism and became fully involved in the European and Pacific theaters of World War II.
U.S. emerges from World War II as the leading global superpower	National Security Act of 1947 lays the foundation for the National Security Council (NSC), Central Intelligence Agency (CIA), Department of Defense (DOD), and other military/foreign policy reorganizations.
Soviets launch Sputnik-1 as first artificial Earth satellite	The quest to put a man on the moon was national policy, the creation of the National Aeronautics and Space Administration (NASA), the National Defense Education Act of 1958.
U.S. military campaigns failed in Iran (1980) and Granada (1983)	Goldwater-Nichols Act of 1986 mandates joint operations as the standard for U.S. military preparedness.
Soviet Union collapses	The intelligence community reshuffled; foreign policy and joint force enterprises reduced in scope/scale.
Attacks of 11 September 2001	Department of Homeland Security (DHS) created, PATRIOT Act enacted, Global War on Terror initiated.

critical questions such as "Why are we competing?" or "Competing over what?" (Ashford 2021). Others took an *ad hominem* approach against the President, with theories ranging from simple predictions of failure (e.g., Trump's domestic and foreign policy agenda are "working against the United States' ability to success-fully engage in [this] kind of great-power strategic competition" [Blankenship and Denison 2019]) to unnecessarily complex origin stories (e.g., great power competition is "pique," not "policy," and stems from a jilted Trump who believes Vladimir Putin and Xi Jinping spurned his transactional diplomacy attempts [Larson 2021, pp. 47–80]). Woven throughout many of these criticisms is the misperception the United States unilaterally (or worse, arbitrarily/vindictively) initiated great power competition – which returns the discussion to the perceived absence of a focusing event.

This chapter argues that both the need for competition and the concept of competition itself are ubiquitously understood across the interagency and society in general – which in turn leads to disjointed policy implementation. As observed by Nobel Laureate Elinor Ostrom, "language and discourses cannot exist if there is no common agreement among some community of users about what is included or excluded by the use of a particular term" (Ostrom 1980, p. 200). Consequently, this paper first establishes a framework clarifying the concept of competition and drawing attention to the actions of Russia and China that precipitated the policy (i.e., the focusing event). This is done through qualitative literature reflections combined with the empirical observations of the author as a defense practitioner–scholar (Section One). Second, this chapter leverages a content analysis dataset derived from publication databases to make quantitative inferences regarding the nature of the broader literature (Section Two). Finally, this chapter advocates for a second generation of research based on the findings of Sections One and Two.

Section One: Qualitative and Empirical Observations on the Essence of Competition

Dissecting the intangible aspects of policy is a challenge, a reality no more succinctly summarized in Justice Potter Stewart's (1965) *Jacobellis v. Ohio* infamous Supreme Court opinion defining *obscenity* as "I know it when I see it." Samuel Huntington (1996, p. 21) overcame portions of this paradox in his seminal work *The Clash of Civilizations* through a juxtaposition of unifying and divisive forces (i.e., "we know who we are only when we know who we are not and often only when we know whom we are against"). The following co-opts this approach by establishing what competition is not with the intent of using the antithetical to clarify the affirmative.

What Competition Is Not …

Negation #1: Competition is not a euphemism reducing all geopolitical tensions to being a prelude to war. When Thucydides (n.d.) expressed that Athens's rising power and the resulting fear in Sparta "made war inevitable," he was a historian analyzing specific circumstances within a specific context – he was not devising a universal model for great power dynamics. Thus, it is important to remember the notion of a "Thucydides's Trap" originated more recently with Graham Allison and the point of emphasis is not causal determinism (i.e., War is inevitable between powers in decline/ascendency) but rather a call to "imaginative statecraft" as an alternative to fatally embracing an arms race (Harvard University n.d.).

Negation #2: Competition is not an endeavor aimed at dissuading adversaries from initiating war – that term exists: deterrence. Although maintaining credible deterrents and managing escalation during a crisis is an aspect of competition, it was also an aspect of U.S. national security policy before the 2017 NSS prioritized great power competition (i.e., if competition equals deterrence then it necessarily follows that a *new* policy of competition is merely a fresh buzzword for an old concept). Deterring aggression is not competition's principal objective and synonymizing the two causes competition to be understood as a military struggle rather than a whole-of-government one. Moreover, when military strategists conflate these terms in practice (e.g., "we can and will deter our competitors in competition, deescalate in crisis, and deny or defeat in conflict" [VanHerck 2021, p. 5]) it increasingly pulls the focus to the Department of Defense (DOD), stunting implementation by biasing discussions in conflict-oriented terminology while simultaneously stifling Interagency coordination.

Negation #3: Competition is not Cold War 2.0. This cannot be overstated, particularly for those recommending that the United States, to borrow a phrase, *pick up where it left off* in 1991 with cookie-cutter policies such as the call to a "Third

Offset" strategy (U.S. Department of Defense 2016). In addition to subliminally reinforcing competition as a preparation for conflict (Negation #1) and a renewed call for innovative deterrence (Negation #2), worse still it implies this chapter in history is a familiar one. The 21st century's technological advancements and forces of globalization (e.g., the internet; cyberspace operations; expanding governmental and commercial access to space; and interconnected markets) make the ongoing great power competition fundamentally different than anything previously experienced.

Negation #4: Competition is not a policy position originating from the U.S. Conversely, the Trump administration simply acknowledged the existence of competition as initiated by (primarily) China and Russia across all instruments of national power – diplomacy, information, military, and economics (DIME). This distinction is paramount to understanding competition holistically because much of the criticism directed at the policy is actually criticism of Trump masquerading as objective commentary. For example, one scholar claimed: "Under Trump, the United States adopted a bullying attitude toward China and an indifferent attitude toward Russia" (Larson 2021, p. 48). Similarly, a *New York Times* writer stated: "Under President Trump, national security concerns over China's rise have *bled into* [emphasis added] trade and economic policy" (Wong 2019, see article's metadata), inferring that Trump was the singular driving force and that the policy struggles under its own weight because he blurred the lines between policy subdisciplines (e.g., national security policy, economic policy). Many of these criticisms are easily reconciled through the notion that competition policy is reactive – reinsert the perceived absence of a focusing event.

What Competition Is ...

Attribute #1: Competition is fundamentally adversarial. Some liken the current competition with China to the 1980 economic competition with Japan (IGCC 2021; Lohr 2011). Although there are undoubtedly similarities within the realm of economic policy, leveraging 1980s Japan as a case study for great power competition in the current geopolitical climate is counter-productive and should be avoided because the economic competition with Japan was not a byproduct of an adversarial nation-against-nation zero-sum game that could feasibly result in war. In simpler terms: Despite the strained relations between Tokyo and Washington (Kim 1999, pp. 143–162), Japan did not want the United States to fail as a society or vice versa. This cannot be said of today's competition. China and Russia are actively advancing their own strategic objectives by challenging core principles of the rules-based international order while the United States is seeking to defend its core national interests and reassure its allies/partners in the process. In short, the adversarial intent to deliberately do harm makes China/Russia's desired end states mutually exclusive from America's.

Attribute #2: Competition takes place below the threshold of armed violence. While this may seem self-evident, the conflict-oriented terminology emerging in the "What Competition is Not ..." negations suggest otherwise. Moreover, given the adversarial intent to do harm (Attribute #1) it must also be recognized that historically the primary mechanism by which states inflict harm is through warfare. As theorized by Prussian General Carl von Clausewitz (1873) – the father of modern military strategy – in his seminal work, *On War*, war is "an act of violence to compel our opponent to fulfill our will;" he further delineated between an immutable nature of war (i.e., armed violence) and an ever-evolving character of war where "violence arms itself with the inventions of Art and Science in order to contend against violence." However, as "Art and Science" evolve, nonviolent technologies become increasingly capable of compelling opponents and/or achieving strategic objectives – thereby blurring the lines between war and peace in what has come to be known as the *gray zone*. This concept will be pursued further in subsequent sections, but for the purposes of Attribute #2 the takeaway is cooperation and conflict exist as dipoles along a geopolitical continuum with competition as an ambiguity between them (Figure 2.1).

Attribute #3: Competition is shaped and informed by the adversary. Much has changed since the Iron Curtain's fall in December 1991, hence why the United States should not consider today's competition to be Cold War 2.0. To be clear, this does not mean that there are no lessons to learn from the Cold War; to the contrary, there are countless. That said, today's strategies must be based on today's players as opposed to archival game film from the past. Ironically, this is precisely how Cold War strategies were originally developed. When George Kennan (writing under the pseudonym "X") advocated for a foreign policy of "containment," he did so after analytically concluding the USSR's ideological, cultural, and socioeconomic foundations "bears within it the seeds of its own decay, and that the sprouting of these seeds is well advanced" (X 1947; U.S. Department of State n.d.). In other words, Kennan recommended containment not to mitigate the damage of Communism spreading (although it did) but rather as a path to victory: Soviet Communism would implode under the weight of its own unsustainable governance, containment would prevent Moscow from extending its lifespan by parasitically draining other nations.

This concept demands a brief parenthetical detour as it brings greater clarity to both Attribute #3 (Competition shaped/informed by the adversary) and by extension Attribute #2 (Competition is below the threshold of armed violence). In fact, the more one understands the adversary, the more competition's existence below the threshold of armed violence gains tractability. While a comprehensive survey of Chinese/Russian activities exceeds the scope of this paper, the most efficient

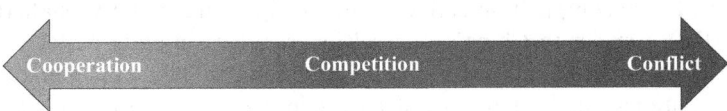

Figure 2.1 The Conflict Continuum.

Source: Author's creation.

mechanism to demonstrate this assertion is key aspects of both countries' military modernizations. This may seem counterintuitive/contradictory after stressing that competition is not a military struggle, yet it is for this reason the argument is compelling.

Containment during the Cold War was bound by the physical constraints of space-time; however, this is not feasible today as space, cyber, interconnected markets, and the information environment make global interactions instantaneous (i.e., as the Clausewitzian "Art and Science" evolve). Thus, it becomes understandable why some contend Russia and China "have pursued their efforts with national-level investments and a singular purpose: to *compete* [emphasis added] with the United States in every [warfighting] domain" (VanHerck 2021, p. 5). Nevertheless, this is another instance where the language "compete" is synonymized with traditional military activities and armed conflict. Although it is not inaccurate to say America's principal competitors have dedicated substantial effort to modernizing their combat capabilities, it is woefully incomplete to portray military modernization as an exclusively combat-oriented endeavor. To the contrary, China and Russia are actively reorganizing and optimizing portions of their armed forces (among other facets of their governments/societies) to specifically engage in the gray zone:

- In 2016, the People's Liberation Army (PLA) created the Strategic Support Forces (SSF) which centralized cyberspace operations, psychological warfare, electronic warfare, and strategic space capabilities under one unified command that reports directly to the Central Military Commission (U.S. Department of Defense 2020, pp. 61–62). Portions of the SSF are explicitly tasked to carry out China's concept of the *Three Warfares* – psychological warfare, public opinion warfare (a.k.a. media warfare), and legal warfare – effectively broadening its military's aperture for competition in the 21st century, "where hearts, minds and opinion are, perhaps, more important than kinetic force projection" (U.S. Department of Defense 2020, p. 62; Halper 2013, p. 31).[2] Serving as a quasi-auxiliary force to the uniformed personnel, China's unofficial network of 50,000–100,000 civilian hackers further enable Beijing to advance its own interests across governmental activities (Hvistendahl 2010).
- Russia tells a similar story with its military intelligence agency, the Main Directorate of the General Staff (a.k.a. the GRU, its original Soviet designation) increasing its operational oversight of espionage, cryptology, unconventional forces (e.g., Spetsnaz), offensive cyberspace operations, influence operations, election interference, assassinations, psychological operations, and other activities designed to accomplish objectives without triggering military retaliation (Bowen 2020). Similar to China's *Three Warfares*, Russia underwrites these aforementioned governmental activities by engaging in "Cyber Diplomacy" within international institutions in order to gain a preemptive advantage by developing new international norms and portraying itself as a paragon of responsible usage of cyber and information tools (Chernenko 2018, pp. 43–50).

In addition to the Ministry of Defense (MOD) personnel, Russia adds a supplementary layer of obfuscation to its gray zone activities by hiring private military corporations (PMCs; which are in and of themselves illegal under Russian domestic law) such as the Wagner Group to secretly carry out the GRU's directives while denying the PMC is on their payroll. Wagner essentially functions as an undeclared branch of the armed forces and thus avoids culpability under the Geneva Convention while propping up Moscow-friendly regimes (e.g., Syria, Libya, Venezuela) and manipulating conditions in support of Russian companies exploiting natural resource deposits (e.g., Sudan, Central African Republic [Chernenko 2018; Warren 2020, pp. 76–77]). Ultimately, the common denominator is these actions are "leveraged by the Kremlin in its military strategy to stall adversaries' responses and make short-term strategic gains" (Lindner 2018).

In light of these actions, the enigmatic focusing event emerges with tangible clarity as one realizes U.S. competition policy is not a unilateral decision but rather reactive. Meanwhile, China/Russia decisively maintain first-mover advantage.

Attribute #4: Finally, competition is a bipartisan national security imperative. The Biden Administration's Interim NSS uses the word "competition" (or a variation thereof) over a dozen times, including a mandate for the Interagency to "develop capabilities to better compete and deter gray zone actions" (U.S. Executive Office of the President 2021b, p. 14). In that same vein, Biden made references to "long-term strategic competition" with China and Russia during his Munich Security Conference remarks in February 2021 (U.S. Executive Office of the President 2021a). America's "adversaries have brought strategic competition to the nation's front door" (Haugh et al. 2020, p. 30) and as such the national security apparatus must think differently about countering these new variants of threats – and thinking differently requires literature that comprehensively accounts for competition policy as interdisciplinary.

Section Two: Quantitative Indicators of Bias in the Great Power Competition Literature

The aforementioned framework for understanding competition is derived from small portions of the literature and the qualitative observations of the author as a defense practitioner. As demonstrated up to this point, there are numerous counterproductive tendencies permeating discussions of great power competition, which for this paper can be summarized as: (1) A tendency to reductively compartmentalize analysis based upon policy sub-disciplines vis-à-vis a unified approach (e.g., "U.S.-Chinese competition primarily concerned trade and technology, the United States–Russian rivalry is in the military-security sphere" (Larson 2021, p. 65); (2) of those sub-disciplines, the dialogue biases heavily towards characterizing competition as a military struggle; and (3) a pervasive tendency to frame competition relative to the Cold War experience.

Data Collection and Automated First-Tier Analysis

Employing Clemson University access to the content analysis tool *Constellate* (in beta at the time of data collection),[3] a unique dataset of 1,390 published documents was aggregated from the JSTOR (820) and Portico (570). These publications (726 articles, 198 chapters, 9 books, 457 reports) were published January 2010–November 2021 and all contain the term "great power competition." The annual distribution of documents is presented in Figure 2.2.

As expected, a surge in publications on this topic emerged after the Trump NSS (December 2017), with 849 documents produced from 2018 to 2021 (61.1%). In Figure 2.2, the 163 documents published in 2021 are included in the overall dataset but omitted intentionally from the graph because the declining publication rate is interpreted as noise (i.e., the spike in publications is significant; the 2021 decline is not). This interpretative normalization is justified based on the assumption the decline likely stems from an erroneous prediction that Biden would not renew Trump's policy (i.e., the topic would no longer be relevant/opportunistically worth researching) and/or the decline is a byproduct of COVID-19's impact on academia.

When the texts of all 1,390 documents are aggregated and placed into a pattern recognition detection script searching for keywords/phrases (up to three words), the word bubble plot of Figure 2.3 materializes without any researcher prompting or specified parameters.

Unsurprisingly, the automated findings synergistically align with Section One's qualitative observations to further support this paper's thesis. For example, combat-oriented terminology such as "military" (averaging 90% or more of documents annually), "security" (88% or more), "conflict" (70% or more), "nuclear" (53% or more), and "defense" (consistently more than 50%) dominating the terminology. Similarly, Cold War references to "Soviet" (consistently more than 50%) occupying similar prevalence as to modern-day competitors such as "Russia" (63% or more) and "Chinese" (57% or more). Notably absent from the word bubble are

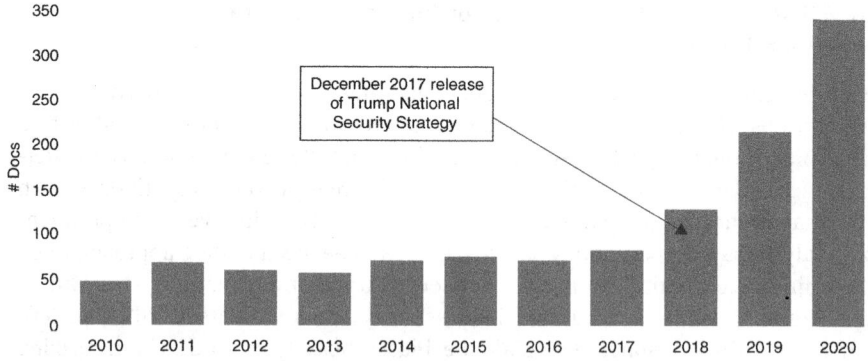

Figure 2.2 Annual Distribution of Publications Containing "Great Power Competition."

Source: Author's creation.

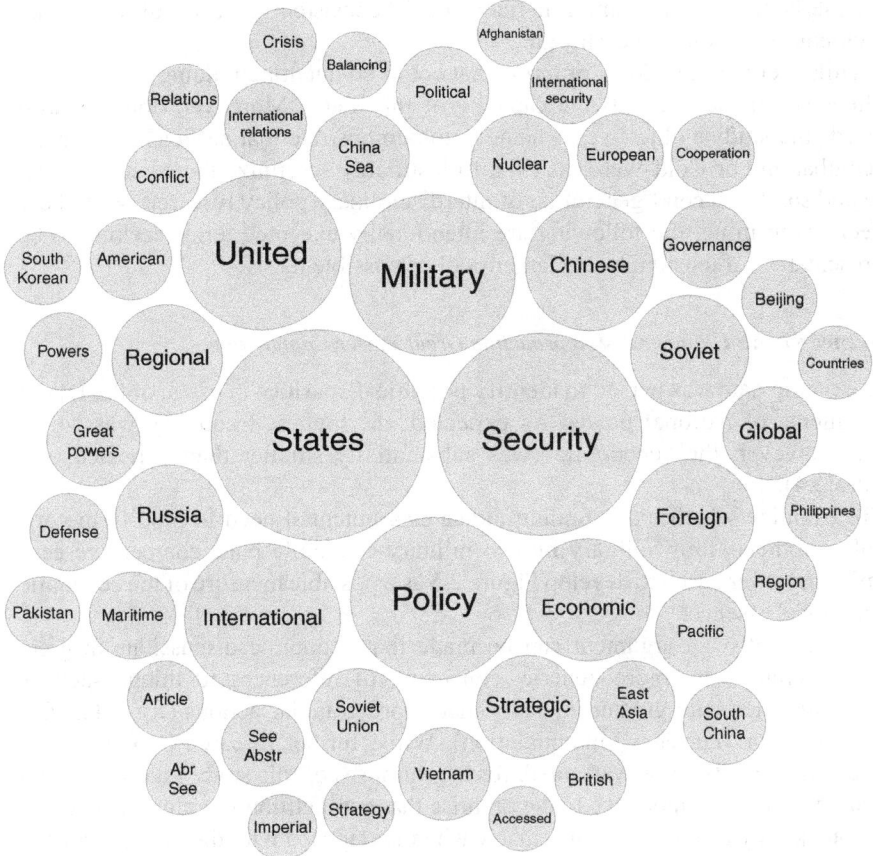

Figure 2.3 Word Bubble Plot of Key Words/Phrases in Great Power Competition Dataset.

Source: Author's creation.

countless terms that one might expect for a more holistic discussion regarding the competition (e.g., misinformation, disinformation, elections, cyber, hack, intellectual property, theft, annex, sanction, tariff, indict).

Methodology: Specific Term Use Frequency

With consistently net-positive findings from the automated parsing, a targeted specific term use frequency methodology was employed to gain increased confidence in both the initial findings and the study's core assertions. The following data is deliberately not depicted in absolute frequency (i.e., the number of appearances in a given year) as that would only replicate the upward curve of Figure 2.1 by correlating with the 2018 spike in literature. Instead, the data visualization is normalized by using relative frequency to demonstrate how the great power competition dialogue is or is not remaining homogenously consistent agnostic of how many documents are published each year.

While there are hundreds of terms that could be quantified, doing so in a comprehensive manner exceeds the capacity of this study. Moreover, doing so also exceeds the study's objectives – namely, to demonstrate that sufficient indicators of combat and/or Cold War bias exist to justifiably scrutinize the existing literature and spark a second-generation of interdisciplinary policy research. With these objectives in mind, the following are intentionally excepted terms seeking to be representative of the admittedly larger pool of possible terms.

Tendency #1: An Unintegrated Approach to Great Power Competition

The existing data was parsed to identify possible disparities in usage of the DIME instruments of national power. As expected, the highest frequency was "military;" however, the "economic" was substantially smaller than expected (see Figure 2.4).

To examine whether this finding indicates a potential need to modify the original hypothesis from military bias to military-economic bias, competitive economic terms were used to develop Figure 2.5 as a possible measure of the economic discussions' essence.

Consequently, an argument can be made that economic discussions in great power literature are more generic, consisting of references to things such as "trade" disputes with greater intensity than more granular actions (e.g., "tariffs," "sanctions," or "currency" manipulation). While further research into economic aspects of competition is warranted, for the purposes of this study Figures 2.4 and 2.5 are considered supportive to the premise that competition literature neglects to integrate policy disciplines (particularly when juxtaposed with the next section).

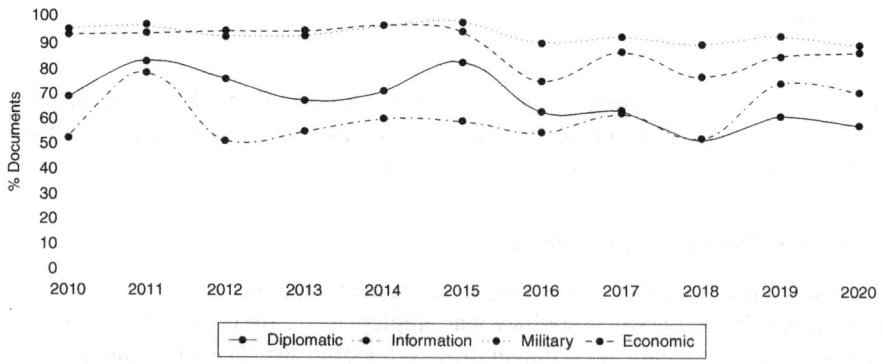

Figure 2.4 Relative Frequency – Diplomatic, Information, Military, Economic.

Source: Author's creation.

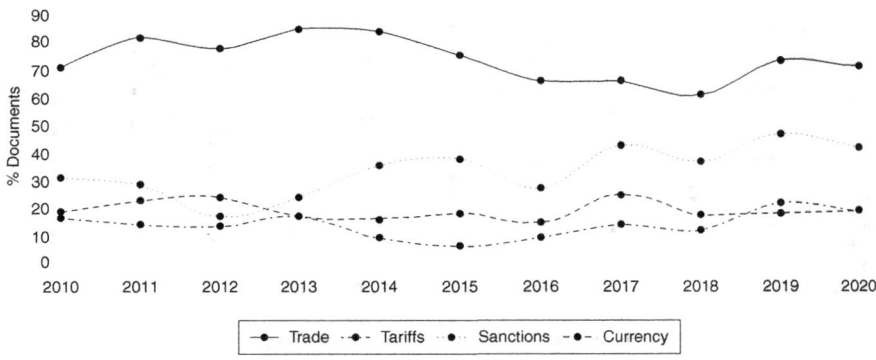

Figure 2.5 Relative Frequency – Trade, Tariffs, Sanctions, Currency.

Source: Author's creation.

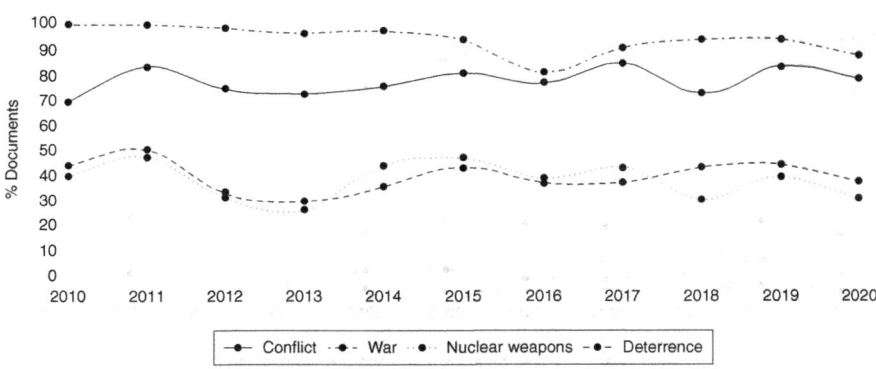

Figure 2.6 Relative Frequency – Conflict, War, Nuclear Weapons, Deterrence.

Source: Author's creation.

Tendency #2: Great Power Competition as a Military Struggle

The frequency distribution of combat-oriented terminology strongly aligns with the possibility of military bias in the literature. In Figure 2.6, not only does "conflict" and "war" dominate the majority of documents (well over 75% of documents contain these terms), but terms such as "nuclear weapons" and "deterrence" – terms only relevant to large-scale combat – appear in roughly half of the documents (an approximate 2:1 ratio of these terms versus some of the economic lexicon in Figure 2.5).

To gather further indications of military-bias in the literature, comparisons were developed based on mechanisms nations utilize to conduct war (see

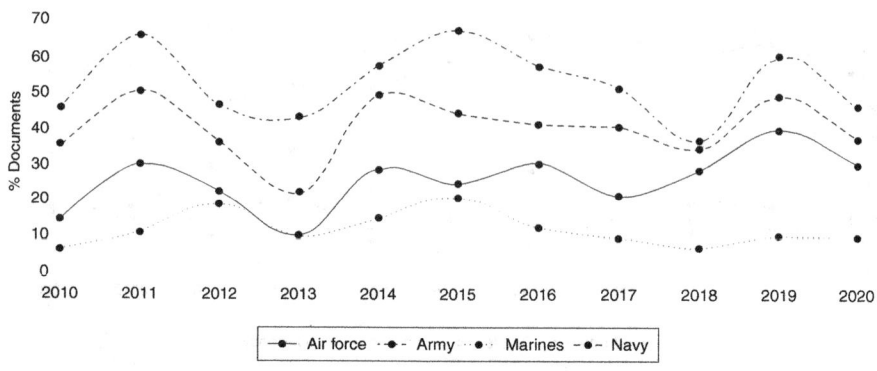

Figure 2.7 Relative Frequency – Air Force, Army, Marines, Navy.

Source: Author's creation.

Note: "Space Force" was intentionally omitted, service branch created 12/20/2019.

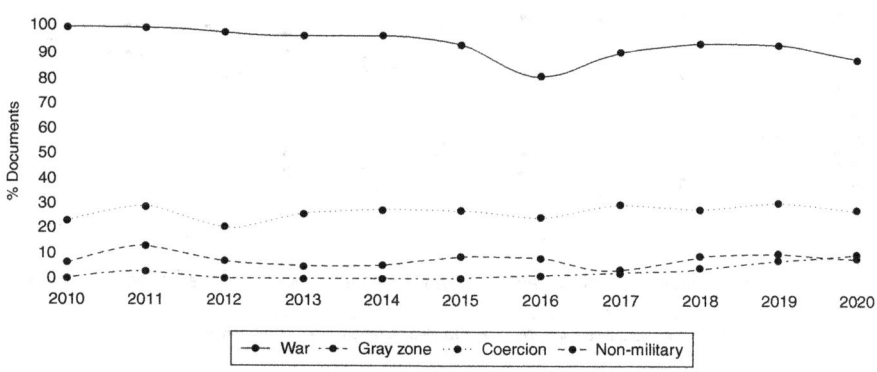

Figure 2.8 Relative Frequency – War, Gray Zone, Coercion, Non Military.

Source: Author's creation.

Figure 2.7's examination of armed services where references to "army" and/ or "navy" emerge in roughly half of the documents). Along these same lines, Figures 2.8–2.12 compare and contrast the disproportionately high use of the term "war" with a range of nonviolent competition actions China/Russia are employing. While there are more terms that could be quantified, these cross-sections are consistent with the researcher's premise.

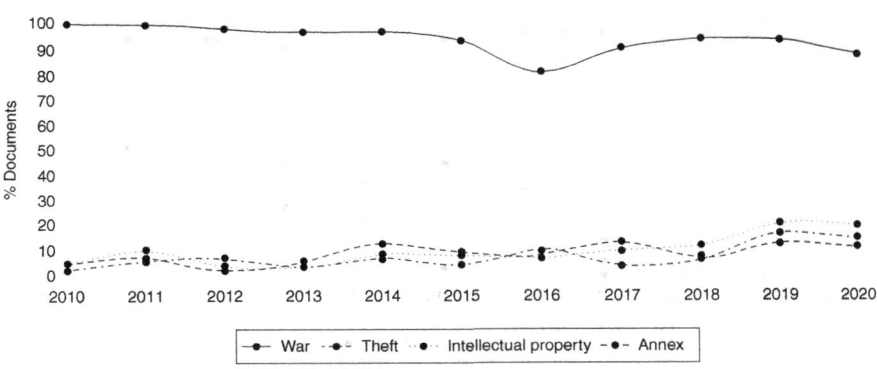

Figure 2.9 Relative Frequency – War, Theft, Intellectual Property, Annex.

Source: Author's creation.

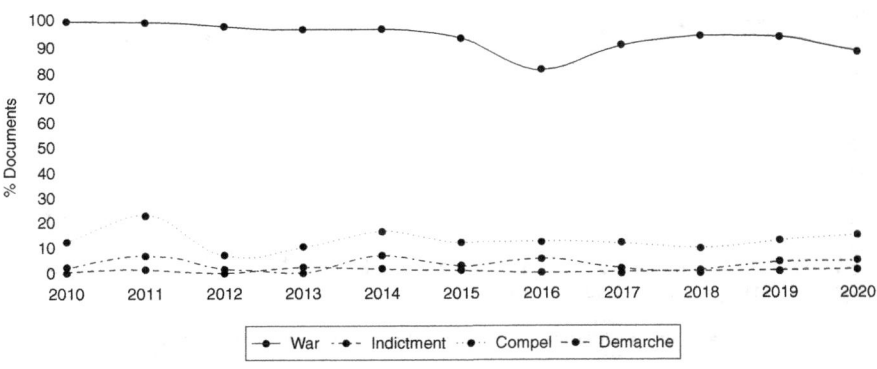

Figure 2.10 Relative Frequency – War, Indictment, Compel, Demarche.

Source: Author's creation.

Tendency #3: Great Power Competition Compared to Cold War Experience

Finally, to measure indicators that competition is being grounded in Cold War experiences, the search parameters are far cleaner than the terminological cross-section developed for Figures 2.6–12 due to less margin for syntax/lexiconic variation. As the old adage goes, if *a picture is worth 1,000 words* then it stands to reason that Figure 2.13 offers a compelling argument for validity – with "Soviet" and/or "Cold War" appearing in well over half of the 1,390 documents and at times appearing proportionately to or even eclipsing the frequency of modern-day competitors "China" and "Russia." For the purposes of this paper/brevity, no further data analysis is pursued along this particular line of inquiry as the general assertion is sufficiently supported by Figure 2.13 enough to reasonably transition to the call for new research.

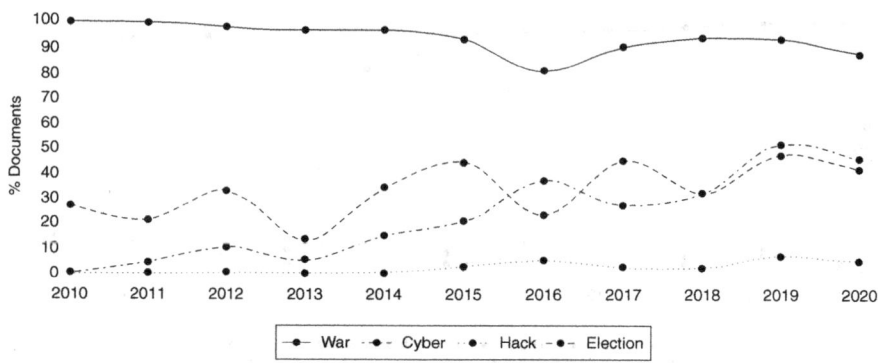

Figure 2.11 Relative Frequency – War, Cyber, Hack, Election.

Source: Author's creation.

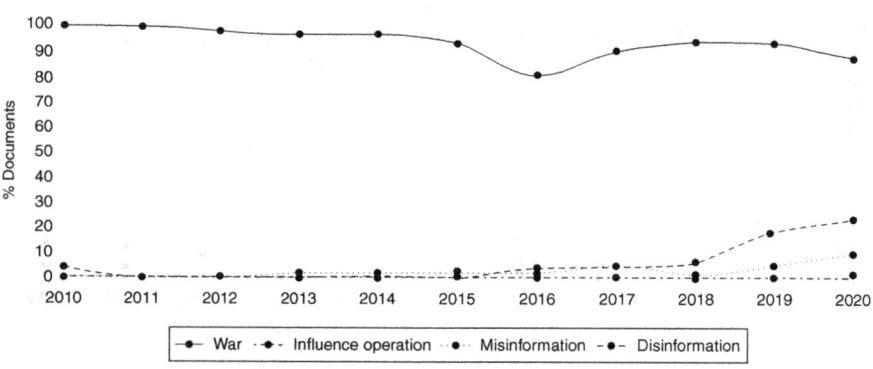

Figure 2.12 Relative Frequency – War, Influence Operation, Misinformation, Disinformation.

Source: Author's creation.

Way Forward: The Need for a Second Generation of Research and Thinking Differently

The research presented above is not an end but a beginning; a foundation upon which to begin a second generation of research on great power competition. Such research and insights from academia and industry are critical for the United States to overcome the first-mover advantage occupied by China and Russia. If the concept of competition is understood as a military struggle analogous to the Cold War experience, the nonmilitary components of the Interagency do not necessarily respond to Biden's Interim NSS for all "departments and agencies to align their actions with this guidance" to engage in competition because they do not relate

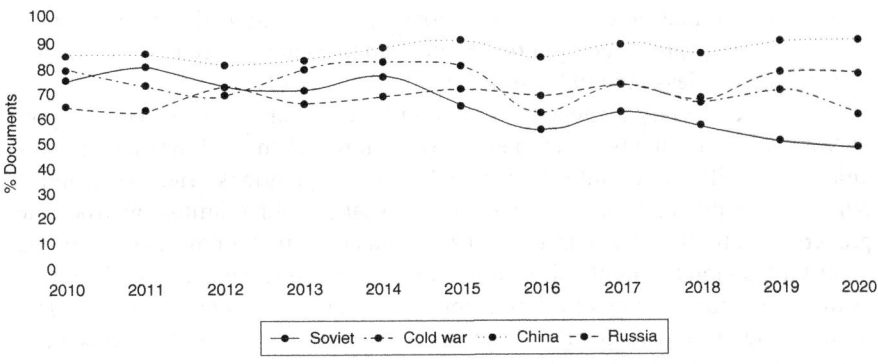

Figure 2.13 Relative Frequency – Soviet, Cold War, China, Russia.

Source: Author's creation.

it to their functional authorities/expertise (U.S. Executive Office of the President 2021a,b, p. 4).

Thus, just as Clausewitz described war as "a continuation of policy by other means" (i.e., by armed violence), so too is competition a continuation of policy by means proving to be misunderstood by many practitioners. Similarly, just as Thucydides (n.d.) contended that nations go to war over *Fear, Honor,* and *Interest,* so too are China and Russia using competition to achieve these same strategic ends – with one key difference. The deliberate pivot to competition is not an altruistic avoidance of war; it is a pragmatic/less-expensive alternative that allows adversaries to achieve the spoils of war without waging war militarily (e.g., the annexation of South China Sea, Crimea, Ukraine [2014–2020];[4] theft of intellectual property; targeting civilian populations with misinformation/disinformation; damaging critical infrastructure). These circumstances demand thinking differently about competition and pursuing new lines of inquiry, such as the following:

- How can diplomatic tools/pressures (e.g., demarche, indictments) be innovatively employed to counter gray zone activities and other nefarious competitive action in support of free societies? By extension, what are the roles of international organizations/collective security frameworks (e.g., NATO, UN, ASEAN) in response to activities below the threshold of armed conflict? What, if any, new allies and partnerships should be established to affirm free societies in response to authoritarian competitive acts that undermine the rule of law and basic human rights? Does domestic and/or international law need to evolve in response to the threat of gray zone actions since collective security invocation requires the use of armed violence?
- In addition to undermining the credibility of free and fair elections, how are adversarial competitors using misinformation/disinformation to achieve strategic objectives and/or protect their own national interests? What are the

similarities/differences in adversaries' tactics, techniques, and procedures within the information environment? How can public education on gray zone actions and great power competition be utilized to create a more informationally resilient society less vulnerable to informational attack?
- How can the military's capacity for action be used outside of war in support of the other instruments of national power? How can the military leverage its unique capabilities for intelligence, cyberspace operations, etc., to generate otherwise unattainable insights and develop targets for elimination from the proverbial battlefield through other instruments of national power (e.g., in lieu of striking a target with physical force, military intelligence data could be used as the legal underpinnings for law enforcement actions with the same net effect of rendering the asset/capability unusable)? Does the concept of deterrence need to broaden beyond the military instrument of national power?
- How can economic instruments (e.g., sanctions, asset freezing) impose costs upon those engaging in destabilizing activities? How can corporate entities at home and abroad (e.g., Google, Meta, Twitter/X, professional sports, Silicon Valley) integrate their economic influence in a whole-of-society response to competitive gray zone actions?

However, pursuing any of these prompts requires thinking differently – especially during instances where an adversarial competitor occupies first-mover advantage. For example, when the Wagner PMC and the Internet Research Agency (IRA) receive contracts and begin taking orders from the MOD for operations outside Russia's sovereign territory, they become an element of the military instrument. As long as they operate in the gray zone, the United States using its armed forces to remove Wagner/IRA off the proverbial battlefield would make the West an *aggressor* who is *escalating* tensions – which by extension only reinforces Moscow's plausible deniability and its strategic advantage. Yet if the West unleashes integrated demarches, indictments, extraditions, criminal dossiers to allied law enforcement, etc. (i.e., Diplomatic Power); public disclosures of previously classified intelligence eliminating plausible deniability (i.e., Informational Power); along with sanctions, asset freezes, etc. (i.e., Economic Power), targets can be neutralized without violence while staying true to liberal democratic values. Moreover, Military Power (particularly its intelligence and cyberspace capabilities) can directly support the other instruments by generating targetable information on gray zone activities.

Conclusion

In conclusion, the aforementioned research prompts and/or anecdotal exemplars are by no means exhaustive and are intended as a starting point to spark innovation in the field of geopolitics and adjacent disciplines. Ultimately, the overarching theme that should be woven throughout future research is the interdisciplinary collaboration across policy genres; instruments of national power are more effective when synchronized. Closing the quantitative gaps in the body of literature (Section Two) will assist in resolving the previously cited misconception that "national security concerns over China's rise have *bled into* [emphasis added] trade and economic policy." Great power competition is not a singular policy type issue and society's understanding of it needs to expand beyond its current state (Section One) in order to effectively counter authoritarian global malign activities in the gray zone and shift towards a proactive strategic policy environment.

Notes

1 More formally defined, incrementalism is: "A model of decision-making in which policy change is accomplished through small, incremental steps that allow decision-makers to adjust policies as they learn from their successes and failures."
2 The Three Warfares are utilized to "to shape international public narratives, weaken the enemy's will, shape diplomatic and political narratives, and advance the PRC's interests through all phases of conflict."
3 See www.constellate.org. Information cutoff point for dataset was November 2021.
4 Although Russia's unlawful war of aggression against Ukraine in 2022 demonstrates that military conflict characterized by armed violence is not a relic of the past, it is critical to understand that even in a traditional conflict the sub-threshold tradecraft perfected during periods of great power competition are deployed alongside their kinetic counterparts. A perfect example is Moscow's nefarious use of troll accounts and fabricated "fact-checking" on the social media platform Telegram during the invasion of Ukraine to attempt to paint Kyiv as immoral propagandists manipulating global public opinion (Silverman and Kao 2022).

References

Ashford, E., 2021. Great Power Competition Is a Recipe for Disaster. *Foreign Policy*, 1 April. Available from: https://foreignpolicy.com/2021/04/01/china-usa-great-power-competition-recipe-for-disaster

Baumgartner, F. R., and Jones, B. D., 2009. *Agendas and Instability in American Politics*. 2nd ed. Chicago: University of Chicago Press.

Birkland, T., 2020. *An Introduction to The Policy Process*. 5th ed. New York: Routledge.

Blankenship, B. D., and Denison, B., 2019. Is America Prepared for Great-Power Competition? *Survival*, 61(5), 43–64. https://doi.org/10.1080/00396 338.2019.1662134

Bowen, A. S., 2020. Russian Military Intelligence: Background and Issues for Congress. *Congressional Research Service*, 24 November. Available from: https://sgp.fas.org/crs/intel/R46616.pdf

Chernenko, E., 2018. Russia's Cyber Diplomacy. *In:* N. Popescu and S. Secrieru, eds. *Hacks, Leaks, and Disruptions: Russian Cyber Strategies*. Paris: European Union Institute for Security Studies, 43–50. Available from: www.jstor.org/stable/resrep21140.8

Clausewitz, C., 1873. *On War*, trans. Graham, J.J. London. Available from: www.clausewitz.com/readings/OnWar1873/BK1ch01.html

Halper, S., 2013. China: The Three Warfares (For Andy Marshall, Director, Office of Net Assessment, Office of the Secretary of Defense). *University of Cambridge*, May. Available from: www.iwp.edu/wp-content/uploads/2019/05/201810171_HalperChinaThreeWarfares.pdf

Harvard University. n.d. Can America and China Escape Thucydides's Trap? *Harvard Kennedy School Belfer Center*. Available from: www.belfercenter.org/thucydides-trap/overview-thucydides-trap

Haugh, T., Hall, N., and Fan, E., 2020. 16th Air Force and Convergence for the Information War. *The Cyber Defense Review*, 5(2), 29–43. Available from: https://cyberdefenserev iew.army.mil/The-Journal/Publications/

Huntington, S. P., 1996. *The Clash of Civilizations and the Remaking of World Order*. New York: Simon & Schuster.

Hvistendahl, M., 2010. China's Hacker Army. *Foreign Policy*, 3 March. Available from: www.foreignpolicy.com/2010/03/03/chinas-hacker-army

Institute on Global Conflict & Cooperation (IGCC), 2021. Virtual Training Course on Geo-Economic and Geostrategic Dimensions of Great Power Competition in the 21st Century. *In: 2021–2022 Hybrid Training Program on Great Power Dynamics*, 10–12 August 2021. San Diego. University of California.

Kim, H.K., 1999. U.S.-Japan Relations: A Global Partnership "In Preparation." *Asian Perspective*, 23(2), 143–162. Available from: www.jstor.org/stable/42704211

Kingdon, J., 2011. *Agendas, Alternatives, and Public Policies*. 2nd ed. Glenview, IL: Longman.

Larson, D. W., 2021. Policy or Pique? Trump and the Turn to Great Power Competition. *Political Science Quarterly*, 136(1), 47–80. https://doi-org.libproxy.clemson.edu/10.1002/polq.13134

Lindner, A., 2018. Russian Private Military Companies in Syria and Beyond. *CSIS's New Perspectives in Foreign Policy* 16, 17 October. Available from: www.csis.org/npfp/russ ian-private-military-companies-syria-and-beyond

Lohr, S., 2011. Maybe Japan was Just a Warm-up to the Rivalry with China. *The New York Times*, 23 January. Available from: www.nytimes.com/2011/01/23/business/23japan.html

Ostrom, E., 1980. Is It B or Not-B? That Is the Question. *Social Science Quarterly*, 61(2), 198–202. Available from: www.jstor.com/stable/42860712

Silverman, C. and Kao, J., 2022. In the Ukraine Conflict, Fake Fact-Checks Are Being Used to Spread Disinformation. *ProPublica*, 8 March. Available from: www.propublica.org/article/in-the-ukraine-conflict-fake-fact-checks-are-being-used-to-spread-disinformation

Stewart, P., 1965. Nico JACOBELLIS, Appellant, v. STATE OF OHIO. *Cornell Law School Legal Information Institute*. www.law.cornell.edu/supremecourt/text/378/184

Thucydides. n.d. *The History of the Peloponnesian War*, trans. Richard Crawley. Project Gutenberg, 7 September 2021. Available from: www.gutenberg.org/files/7142/7142-h/7142-h.htm

U.S. Department of Defense, 2016. Office of the Deputy Security of Defense. *Remarks by Deputy Secretary Work on Third Offset Strategy*. 28 April. Available from: www.defe nse.gov/Newsroom/Speeches/Speech/Article/753482/remarks-by-d%20eputy-secretary-work-on-third-offset-strategy

U.S. Department of Defense, 2018. Office of the Secretary of Defense. *2018 National Defense Strategy*. Available from: https://dod.defense.gov/Portals/1/Documents/pubs/2018-National-Defense-Strategy-Summary.pdf

U.S. Department of Defense, 2020. Office of the Secretary of Defense. *Military and Security Developments Involve the People's Republic of China: 2020 Annual Report to Congress*. 21 August. Available from: https://media.defense.gov/2020/Sep/01/2002488689/-1/-1/1/2020-DOD-CHINA-MILITARY-POWER-REPORT-FINAL.PDF

U.S. Department of State. n.d. Office of the Historian. Foreign Service Institute. *George Kennan and Containment*. Available from: https://history.state.gov/departmenthistory/short-history/kennan

U.S. Executive Office of the President. 2015. National Security Council. *2015 National Security Strategy*, February. https://obamawhitehouse.archives.gov/sites/default/files/docs/2015_national_security_strategy_2.pdf

U.S. Executive Office of the President, 2017. *2017 National Security Strategy*. Available from: https://trumpwhitehouse.archives.gov/wp-content/uploads/2017/12/NSS-Final-12-18-2017-0905.pdf

U.S. Executive Office of the President, 2021a. White House Briefing Room. *Remarks by President Biden at the 2021 Virtual Munich Security Conference*, 19 February. Available from: www.whitehouse.gov/briefing-room/speeches-remarks/2021/02/19/remarks-by-president-biden-at-the-2021-virtual-munich-security-conference/

U.S. Executive Office of the President, 2021b. National Security Council. *Interim National Security Strategic Guidance*, March. Available from: www.whitehouse.gov/wp-content/uploads/2021/03/NSC-1v2.pdf

VanHerck, G. D., 2021. Deter in Competition, Deescalate in Crisis, and Defeat in Conflict. *Joint Force Quarterly*, 101(2), 4–10. Available from: www.norad.mil/Newsroom/Article/2557947/deter-in-competition-deescalate-in-crisis-and-defeat-in-conflict

Warren, J., 2020. Not All Wars are Violent: Identifying Faulty Assumptions for the Information War. *Air & Space Power Journal*, 34(4), 75–90. Available from: www.airuniversity.af.edu/Portals/10/ASPJ/journals/Volume-34_Issue-4/F-Warren.pdf

Wong, E., 2019. U.S. Versus China: A New Era of Great Power Competition, but Without Boundaries. *New York Times*, 26 June. Available from: www.nytimes.com/2019/06/26/world/asia/united-states-china-conflict.html

X (George Kennan), 1947. The Sources of Soviet Conduct. *Foreign Affairs*, July. Available from: www.foreignaffairs.com/articles/russian-federation/1947-07-01/sources-soviet-conduct

3 Russian New Generation Warfare in the Baltic States and Beyond

Sandor Fabian and Janis Berzins

The role of nonmilitary means of achieving political and strategic goals has grown, and, in many cases, they have exceeded the power of force of weapons in their effectiveness.[1]

(General Valery Gerasimov 2016)

Introduction

As a result of Russia's 2008 invasion of Georgia, its illegal annexation of Crimea in 2014, and the Russian full-scale (re)invasion of Ukraine in 2022, the world has been witnessing the re-emergence of an aggressive, revisionist Russia. The recent Russian operations have shown that its strategy is highly adaptable and includes both military and nonmilitary instruments tailored to a given theater's strategic necessities. In this chapter, we analyze Russia's nonmilitary instruments and its strategic approach toward the West which seeks to achieve Russia's political aims using influence operations and the exploitation of cyberspace. The scope of our analysis focuses on the period between the annexation of Crimea and the separatist conflict in Eastern Ukraine before the invasion and transition to the current conventional warfare. We argue that the challenge presented by Russia's strategic ambitions and its willingness to aggressively challenge the current world order using its new generation warfare (NGW) concept (Bērziņš 2020) (alongside a similar Chinese approach, so-called Unrestricted Warfare) has returned the world to an era of great power competition (Mattsson 2015).

While the 2018 U.S. National Security Strategy and the NATO 2030 Strategic Concept clearly recognize that the West is, and will be, in a perpetual competition with near-peer competitors for decades to come, some critics argue that neither the U.S. government nor its allies and partners are paying sufficient attention to the focal points of their competitors' strategies: information operations and malicious exploitation of the cyber domain. Despite recent steps taken by the U.S. Department of Defense and various NATO bodies, the overwhelming focus of Western strategy remains building a more lethal, kinetic force (Mattis 2018). This chapter leverages analysis of the United States and Russian competition in the Baltic states and some other examples in order to assess the relative efficacy of these distinct approaches.

DOI: 10.4324/9781003425304-4

Kinetic Approaches in the Baltics

Since regaining their independence from the Soviet Union in 1990, the three Baltic states – Latvia, Lithuania, and Estonia – have been under constant Russian pressure (Lane 2014). One way this pressure has manifested itself is through the conventional military threat, punctuated by ceaseless airspace violations by Russian military aircraft, unannounced, large-scale military exercises along their borders, and the general disposition of Russian military forces in the Baltic's Western-most military district. To assess the seriousness of the conventional military threat and to explore the likely outcome of a Russian invasion of the three Baltic states, a series of wargames were conducted in 2014 and 2015. The results of these exercises showed that, given the military capabilities of the Baltic states and NATO's force posture at the time, Russian forces could overrun the Baltic states in 36–60 hours (Shlapak and Johnson 2016). Moreover, effectively deterring a Russian conventional attack would require permanent deployment to the Baltics of at least seven NATO brigades, including three heavy brigades, as well as appropriate air support and other enablers. Although the design of these wargames had many limitations, including faulty assumptions, ignorance of the terrain, and exclusion of newly developed capabilities their results seem to justify the current Western strategic approach and its continued focus on conventional kinetic capabilities.

Accordingly, NATO has deployed a multinational battalion battle group to each Baltic state and Poland as part of its NATO Enhanced Forward Presence concept, boosted the capabilities and readiness levels of the NATO Response Force, and the U.S. unilaterally forward deployed heavy forces into the region on a rotational basis. For their part, the Baltic states have significantly increased defense spending and adopted new, total defense strategies that emphasize, whole-of-society resistance against any Russian military incursion (Kojala and Keršanskas 2015). These decisions have fundamentally shifted the strategic landscape in the region, such that the probability of the Baltic states becoming targets of Russian conventional military action has become very small (Galeotti 2019). Beyond the sub-optimal strategic landscape for pursuing traditional military action, additional factors suggest that Russia has switched its attention toward more novel approaches when it comes to the Baltic states (Ploumis 2022).

Russia's New Approach to Conflict

Following the end of the Cold War, Russia faced significant resource constraints, impinging its ability to maintain its military research and development competition with the United States and the West. Despite limited resources, Russia's global ambitions did not diminish (Ploumis 2022).

While Russia's political elite seemingly gave up on the idea of re-creating the Soviet Union by force, they never abandoned the goal of reinvigorating Russia's status as a global power. To achieve this goal, Russia has been challenging American unilateralism, trying to establish spheres of influence, creating strategic depth, and changing current global security and defense frameworks through the application of political, economic, cultural, psychological, and other non-kinetic tools. The newfound emphasis on asymmetric approaches to competition is the result of a long process.

During the 1990s, Russian experts spent more time and effort on developing alternative concepts, including their theory of information operations, than their Western peers. However, Western doctrine, especially American military doctrine, continued to shape the general direction of Russian military thinking (Bērziņš 2020). As a result, three distinct waves of Russian military literature emerged, detailing its approach to conflict. American literature on low-intensity conflicts and Russian observations of the same phenomena influenced the way that the Russian military literature analyzed conflict through the mid-1990s (Mattsson 2015). These factors shaped the Russian view of how to conduct operations below the threshold of conventional armed conflict. From the mid-1990s through the early 2000s, the principles and lessons of Operation DESERT STORM and NATO's air campaign in Yugoslavia inspired Russian military strategists, notably General-Major Vladimir Slipchenko's theory of Sixth Generation Warfare or "Contactless War" (Mattsson 2015). Ultimately, the observation of Western conventional forces' technological overmatch and their ability to destroy enemy targets from great distances with highly precise weapons in these conflicts made Russian strategists realize that it would be extremely difficult to overcome such advantages through conventional means. Consequently, attention was redirected toward unconventional thinking and concepts. In the 2000s, the American Network Centric Warfare doctrine inspired similar developments in Russian military thinking (Office of Force Transformation 2005). Starting in 2008, Russian military experts, most notably Sergey Chekinov and Sergey Bogdanov, amalgamated these concepts into a single approach: NGW (Danyk et al. 2017). As a concept, NGW approaches conflict as a continuum with varying intensity and changing centers of gravity. The shaping of the operational environment, the creation of favorable conditions, and preparations for conventional war are viewed as forever ongoing efforts (Göransson 2022). Continuous activities are executed to weaken, isolate, and destabilize target societies followed by more conventional military actions if necessary. According to Bērziņš (2020), Russian NGW can actually be divided into eight clearly distinguishable phases.

The first phase includes psychological, informational, diplomatic, and economic measures to create favorable conditions for the follow-on actions. The second phase focuses on the deception and confusion of political and military leadership of the targeted country through coordinated activities, including the leaking of sensitive information or spreading fake data. The next phase includes bribing, blackmailing, and intimidating key leaders to make them not to fulfill their duties appropriately during a crisis. The fourth phase focuses on the creation of discontent and confusion between the leadership and the population of the targeted country through destabilizing propaganda supported by covert Russian subversive activities. The fifth phase is about shaping the environment for an upcoming physical action through the creation of no-fly zones and physical blockades of main supply lines.

The next phase focuses on actual military action, including the coordinated and de-conflicted operations of conventional formations, special operations forces, secret services, and private military contractor companies. During the seventh phase, we should see the continuation of information operations augmented by continued military operations and deception. The eighth phase includes the physical occupation of the last points of resistance and, if necessary, the physical destruction of the enemy units (Bērziņš 2020). Russian Operations in the Crimea and Eastern Ukraine relied heavily on these phases and the basic principles of NGW. Starting in 2016, Russian military strategists replaced NGW with the term Hybrid Warfare to refer to the employment of Color Revolutions and political warfare by the West, especially the United States. In other words, in their own view Russia does not engage in Hybrid Warfare, but the West does (Mitchell 2012).

Influence, psychological, and information operations have been part of Russian doctrine since the early days of the Soviet Union, but historically they have been relegated to supporting roles. Recent Russian military thinking has observed that non-kinetic instruments will become dominant in future warfare. This realization is now clearly reflected in most current Russian policy and doctrinal documents. However, while the U.S. Joint Publication 3–13 Information Operations presents a single, overarching information operations doctrine, no Russian parallel exists until today (United States Department of Defense 2012). Yet, the 2000 Russian Information Security Doctrine, 2011 Russian Cyber-Warfare Strategy, and 2014 Russian Military Doctrine all discuss aspects of the goals, importance, nature, and characteristics of information operations and how they might fit into overarching Russian strategic thinking (Fabian 2019). Bazylev et al. (2012) summarize the objectives of Russian information operations as being to disorganize (disrupt) key military, industrial, and administrative capabilities in target states, as well as to bring informational-psychological pressure to bear on the adversary through the use of state-of-the-art information technologies and assets including sophisticated computer programs, nuclear technology, and social media bots. As Blank (2017) observed, while Russian and Western tools for information operations are similar, their employment varies considerably. In particular, Russia, to a much greater extent, sees information operations playing a leading role in its confrontation with the West. Concomitantly, it routinely employs a strategically planned onslaught of disinformation and propaganda designed to manipulate public opinion in target states.

Russian Organizations Involved in Information Operations

Within the Russian federal system several organizations are included in the handling of information operations capabilities. The four most important organizations are the Federal Security Service (FSB), the Foreign Intelligence Service (SVR), the Federal Protection Service (FSO), and the military intelligence (GRU). While the first three were created from the remains of the KGB and directly answer to the Russian president, the GRU is subordinated to the Ministry of Defense. Another organization, the Federal Agency of Government Communication and Information, also existed in Russia between 1993 and 2004, with about 54,000 employees (Fabian 2019). This organization was similar to the U.S. National Security Agency and was responsible for special communications, cryptographic security, technical intelligence, counterintelligence, codebreaking and telecommunication and information protection but was dismantled in April 2003 due to corruption accusations. Its roles, responsibilities and resources were distributed among the above-described four agencies (Boyko et al. 2009).

The largest organization is the FSB with about 270,000 employees (Heickerö 2010). As its name suggests, FSB focuses on state security through law-enforcement-related activities and counterintelligence. The 1995 Law on Operative Search and Seizures provides authorization for FSB to conduct Internet surveillance, wiretapping and opening private mails. The law also enables FSB personnel to trespass private residence without formal court authorization. The FSB also inherited the "Service A Directorate" from KGB and with that responsible for the execution of deception (*maskirovka*). Under the Cold War era name "active measures" FSB co-ordinates the dissemination of false information mainly through mass media and the internet with the aim to influence the adversaries' population (Heickerö 2010).

The SVR, alongside with GRU, is responsible for collecting foreign intelligence in the topics of economy, technology, defense, etc., to support state-level decision-making. Although SVR's roughly 11,800-person organization primarily focuses on human intelligence it also possesses the capabilities to conduct signal intelligence and wiretapping both on fixed and wireless lines. The last organization that is directly subordinated to the Russian president is the FSO. With its approximately 20,000 employees FSO is responsible for supporting top-level communication through the execution of strategic signal intelligence. It conducts surveillance of the internet, fixed, and wireless communications and even satellites (Heickerö 2010).

The fourth major federal organization that has a significant stake in information operations is the military intelligence agency, GRU. This organization's primary focus is on military-related information and gathers it mostly through military attachés, and foreign agents. The GRU utilizes all means of intelligence and surveillance including open-source intelligence, signal intelligence, imagery intelligence, etc. The GRU also has specialized military units under its command mostly responsible for electronic intelligence (Heickerö 2010).

This short analysis of the roles and responsibilities of the different organizations involved in Russian information operations seems to highlight another potential challenge for those who are advocating for the existence of a single, coherent, Russian hybrid warfare strategy. These organizations once again, do not seem to be extremely different from their American counterparts. Additionally, the complexity and multiplicity of these organizations seems at the least obstruct, or at the most, prevent the development and employment of an overarching strategic framework. This begs the question whether the other key factor, the weaponization of the cyberspace points to the same direction or it finally provides some support to the existence of a comprehensive Russian hybrid warfare strategy.

Russia in Cyberspace

The concepts of the cyber domain and cyber warfare are Western phenomena and Russian sources are more likely to refer to the concept of "information space" rather than "cyber space" (Mancuso 2023). For Russia, the cyber domain seems to be a new dimension "in which to attack enemy centers of gravity and critical vulnerabilities and break the enemy's resistance." Chekinov and Bogdanov (2015) seem to explain how Russia understands the cyber domain by suggesting that

> information warfare in the new conditions will be the starting point of every action now called the new type of warfare, or hybrid war, in which broad use will be made of the mass media and, where feasible, global computer networks (blogs, various social networks, and other resources.
>
> (Chekinov and Bogdanov 2015, p. 44)

Furthermore, Nikolai Kuryanovich, a former Russian Duma member, also argued that

> in the very near future, many conflicts will not take place on the open field of battle, but rather in spaces on the Internet, fought with the aid of information soldiers… [A] small force of hackers is stronger than the multi-thousand force of the current armed forces.
>
> (Korns and Kastenberg 2009, p. 60)

However, one must assess whether these Russian unofficial assertions have been reflected in actual Russian activities.

Although cyber activities are still quite easy to hide and tracing them back to their initiators most of the time is a mission impossible, Russia has still been accused many times with malicious cyber actions over recent years (Mancuso 2023). These activities included such tactics, techniques, and procedures that demonstrated significant overlap between cyber activism, cybercrime, and government organizations that executed cyber-attacks. Klimburg (2011) argues that

> The differences between these categories of cyber activity are often razor thin, or only in the eye of the beholder. From the perspective of a cyber warrior, cybercrime can offer the technical basis (software tools and logistic support) and cyber terrorism the social basis (personal networks and motivation) with which to execute attacks on the computer networks of enemy groups or nations.
>
> (Klimburg 2011, p. 46)

The overlap discussed by Klimburg (2011) basically suggests that besides its own governmental capabilities the Russian leadership can utilize individuals (mostly nationalistic hackers – hacktivists) and criminal networks easily for its purposes while maintaining plausible deniability.

Although it has not been confirmed, it looks like the Russian government relies on criminal networks to attack the websites of foreign banks, multinational companies as well as other commercial websites. According to Heickerö (2010), as long as these organizations do not target Russian interests, they are not only allowed by the government to operate freely but sometimes are even contracted to do the "dirty work" for government agencies (Heickerö 2010). As another potential puzzle, Flook (2009) suggests that the level of coordination, sophistication, target-selection, and the timing of the attacks conducted by these organizations clearly suggest government involvement in these activities.

Another potential group of individuals whom the Russian government can use for its purposes are nationalistic hackers. These individuals can be utilized for many different activities. Among many other activities they can conduct Distributed Denial of Service (DDoS) attacks, stealing and leaking emails, hacking into news outlets to modify information, or deny access to specific stories, trolling, and social engineering. We have seen many examples of each of these activities over the recent years both in peacetime including the cases of Estonia and Sweden; and during conflicts in Georgia and Ukraine.

Although it has never been confirmed that Russia was behind the 2007 cyber-attacks targeting Estonia, the international community has accepted that Russia was the perpetrator. During the spring of 2007, Estonia experienced two waves of cyber-attacks. The first wave started at the end of April and peaked on 3 May 2007, while the second one peaked around 9 May 2007 (Heickerö 2010). During the first phase Russian "independent" hackers put out calls on social media sites and recruited nationalistic individuals to conduct attacks against Estonian websites. They provided the tools and detailed instructions online and volunteers could easily download the "cyber weapons." The primary targets were the Estonian Ministry of Defence's website, the Government Briefing Room's system and the websites of the largest Estonian parties (Heickerö 2010). Due to lack of critical mass these attacks did not cause significant damage. The second wave was more sophisticated and led to more serious results. This time large botnets were used to conduct DDoS attacks with the aim to overwhelm information flow to key Estonian systems. During these attacks more than one million computers linked into more than 20,000 networks were affected. Although when tried to trace back the IP addresses of the attacking computers, the investigators initially learned that they were located in around 178 countries, and further analysis suggested the serious involvement of Russian individuals (at this time without any indications of Russian government involvement). Individuals connected to the Nashi Youth group[2] later claimed that they were behind the attacks.

Cyber-attacks can have multiple goals. While in the case of Estonia the actions were clearly aimed at causing damage, they might only serve demonstrative purposes. In 2014, when Sweden seriously considered joining NATO, Russia not only harassed the Swedish government in the physical space but seemingly conducted several cyber actions as well to deter the country joining NATO.

According to an unpublished study, it is possible that "demonstrative actions in cyberspace that inflict no substantial economic or other damage to the target state [but only] cause panic among the population and, as a result, distrust in the authorities."[3] Swedish government officials seem to think that this possibility might be the explanation for multiple unexpected and unexplained cyber incidences including the disruption of the Swedish air traffic control system and problems with several Swedish media outlets' websites due to DDoS attacks, that happened during the period of this intensified tension. While these peacetime activities remain the same during wartime, as the Georgian and Ukrainian example suggest, they are augmented with some additional actions. Both the 2008 Georgian War and the 2014 annexation of Crimea demonstrated how kinetic military actions can be augmented with actions conducted in the digital domain.

On 8 August 2008, at the same time when Russian military forces started their offensive operations, multiple cyber-attacks were also launched against high-visibility Georgian targets, including the websites of the Georgian Parliament, the President, the Ministry of Defense, the Ministry of Foreign Affairs, the National Bank, and the major news outlets in Georgia (Heickerö 2010). DDoS attacks and website defacements were the most common attacks executed by multiple botnets. Georgian President Mikheil Saakashvili's personal website was manipulated by hackers who posted pictures of Adolf Hitler on the presidential website and compared him to the Georgian president (Heickerö 2010). These types of attacks kept intensifying as the fighting became more intense between the two sides. Some of the methods used in the Georgian case were very similar to those measures used against Estonia, while some new ones were also introduced. Private citizens utilized social media sites to recruit and coordinate volunteers to conduct malicious activities. Later investigations identified two Russian hacker fora, stopgeorgia.ru, and Xakep.ru, as potential originators and coordinators of the cyber activity that supported the physical actions of the military. As a new method, attackers utilized so-called SQL injections, a specific computer code that confuses the targeted website's back-end database (Heickerö 2010). Millions of such injections seem to have been used in the Georgian conflict leading to many important servers to become not operational. These types of attacks had significant psychological effects since in many cases these servers were used in media and communication outlets, and their compromise hindered the government's ability to effectively communicate with the population.

Once again, the involvement of the Russian government has never been proved beyond doubt, but several characteristics of the activities suggest governmental connections. These characteristics include the unique timing of the cyber-attacks, their target selection, the content of the attack against the president's website, and the fact that the IP addresses associated with the hosting firm (Steadyhost) used by stopgeorgia.ru are located in the same building as one of the Russian Ministry of Defense's research institute (which also happen to be situated on the same street with the GRU headquarters) (Heickerö 2010).

In the Ukrainian case, Russia seems to have used government organizations and some non-attributable entities in its cyber activities even prior to the actual annexation of Crimea in 2014. Ukrainian agencies dealing with cyber security reported the presence of malicious codes called "Uroburos" on Ukrainian official computers as early as 2010. Later, these codes were connected to a large-scale Russian cyber campaign code-named "Snake" that was launched against multiple European countries with the main goal of gathering information about government communications. These codes also had the capability to take control of computers and shut down information systems (Heickerö 2010).

During the actual conflict beyond the now well-known DDoS attacks by "friendly" hackers and botnets, Russia also used Special Forces on the ground with mobile, man-portable equipment to monitor and disrupt cellular services in Crimea and surrounding areas. Additionally, as the Ukrainian mobile service provider Ukrtelecom later announced, its facilities were quickly taken over by individuals who were intentionally looking for, and tempering with, the fiber optic cables of the buildings. Later, these Crimean facilities were traced back as the points of origin of DDoS attacks that targeted the cell phones of Ukrainian politicians in Kyiv.

Russian Information Operations in the Baltics

The strategic goal of Russian information operations in the Baltic states is to create distance between the Baltics and the West. The Russian approach seeks to achieve strategic ends in the Baltics through influence operations rather than conventional kinetic means. The Russian concept is rooted in the idea that democratic societies are vulnerable to political manipulation, and exploiting this perceived weakness is far less costly than pursuing annexation or occupation. Consequently, Russian information operations in the Baltics focus on nine objectives (Nagorny and Shurygin 2013).

First, Russia seeks to encourage and support armed actions by separatist groups with the objective of promoting chaos and territorial disintegration. Second, increase polarization between the elite and general society of the Baltic states to foment a crisis of values followed by a process of orientation toward Western values. Third, Russia intends to demoralize the Baltic states' military and otherwise attrite their resolve. The fourth objective is to undermine socio-economic stability and engender socio-political crises. Next, Russia also seeks to intensify simultaneous forms and models of psychological warfare to demoralize the Baltic population and break their resolve. Seventh, Russia tries to incite mass panic and degrade confidence in key government institutions. The eighth objective is to defame political leaders not aligned with Russian interests and, finally, to undermine international alliances and partnerships of the Baltic states (Nagorny and Shurygin 2013).

The Russian concept leverages Soviet-era thinking on Reflexive Control (Kowalewski 2017). Specifically, Russia must obtain a detailed understanding of extant fissures in Baltic societies and use this knowledge to convey to an opponent specifically prepared information to incline him/her to voluntarily make the predetermined decision desired by Russia. Recognizing resistance to deeper ties with Russia in the Baltic states, Russia has opted to diversify its messaging beyond simply pro-Russia and pro-Eurasia content. Instead, content attempts to convince the population that their current alignment with the West, embrace of democracy, and membership in NATO and the EU are in some way deleterious. These information campaigns often seek to highlight how the innate moral values of Baltic populations are inherently different from Western values (and not coincidentally more similar to the traditional values championed by Russia). Russian and local language traditional media, as well as social media and the internet are heavily leveraged by Russia toward this end. The main narratives broadcast through these channels include arguments such as (Vasu et al. 2018):

- Russian-speaking minorities are marginalized and treated unfairly by the government.
- The Baltic states are weak states where corruption is widespread;
- EU membership resulted in economic and social underdevelopment of Baltic states, which should therefore follow their own path without foreign interference.
- EU membership is equivalent to being in the Soviet Union.
- NATO membership decreases the overall level of security because of possible Russian countermeasures.
- Western values are corrupted. Tolerance toward homosexuals and other minorities is presented as the moral degradation of traditional family values.
- There is no real democracy in the West. Politicians are puppets controlled by the financial system to work against the real interests of the population.
- Fascism is glorified in the Baltic states.

Some concrete examples of Russian activities include, but are not limited to:

- A Russian plot to support a pro-Russian coalition to dominate the Riga City Council in 2009 and obtain a parliamentary majority during the 2010 elections.
- Russian attempts to undermine initiatives to fully integrate Russian-speaking Latvians into society. These attempts were advanced under the guise of supporting and protecting Russian speakers within the *Russkiy Mir* (Russian World) initiative.
- Pro-Russia pseudo-activists' pursuit of Russian language public education, adopting Russian as the second official language in Latvia, Latgale's autonomy, and the morals and family initiative.
- The passportization of Latvia's population, especially its Russian-speaking population.
- Efforts to collaborate with governments to renovate Soviet military memorials.
- Efforts to discredit NATO forces stationed in the Baltic states and Poland (Fabian and Berzins 2021).

Rather than pursue occupation or annexation, Russian information operations seek to manipulate Baltic populations into electing local, populist leaders that either distance the Baltics from the West and/or pursue closer relations with Russia.

Conclusion

Great power competition is here to stay. Although this competition is between major powers, its primary manifestations will continue to occur in small countries of strategically important regions. Moreover, there exist major conceptual differences between the United States and its competitors when it comes to competition tactics. While the United States remains focused on building a more lethal conventional force, its competitors, and especially Russia, have emphasized weaponizing information to achieve their desired end states without firing a shot.

The chapter does not argue against maintaining the West's kinetic capabilities and heightened readiness, but the focus on conventional military power must be complemented with investments that enable the West also effectively to compete with Russia well before any kinetic conflict. Even though the West has devoted efforts toward integrating operations across all domains (including information and cyber) during kinetic engagements, it has not made nearly the same progress with respect to developing approaches for employing information operations during non-kinetic phases of competition. Continued failure to do so will enable Russia to achieve strategic objectives in the Baltics and beyond before the West has an opportunity to engage militarily. In short, a rebalancing between building a more lethal kinetic force for armed conflict and developing a comprehensive strategy and effective "toolbox" for information operations and other non-kinetic forms of competition is critically needed.

Notes

1 Originally published in Military-Industrial Kurier, 27 February 2013. Translated from Russian 21 June 2014 by Robert Coalson, editor, Central News, Radio Free Europe/Radio Liberty.
2 120,000 strong youth organization was established in 2005 originally with the aim to stamp out Nazi sentiment. The group conducts activities both in the physical and the cyber space, and has been accused of spying on, harassing, and attacking its opponents in both domains.
3 For an overview, see Lucas (2016) and Piotrowski (2016).

References

Bazylev, S. I., Dylevsky, I. N., Komov, S. A. and Petrunin, A. N., 2012. The Russian Armed Forces in the Information Environment: Principles, Rules, and Confidence-Building Measures. *Military Thought*, 21(2), 10.

Bērziņš, J., 2020. The Theory and Practice of New Generation Warfare: The Case of Ukraine and Syria. *The Journal of Slavic Military Studies*, 33(3), 355–380. https://doi.org/10.1080/13518046.2020.1824109

Blank, S., 2017. Cyber War and Information War à la Russe. *In:* G. Perkovich and A. E. Levite, eds. *Understanding Cyber Conflict: Fourteen Analogies*. Washington, DC: Georgetown University Press, 1–18. Available from: https://carnegieendowment.org/files/GUP_Perkovich_Levite_UnderstandingCyberConflict_FullText.pdf

Boyko, S. M, Dylevsky, I.N., Komov, S.A., Korotkov, S.V., Rodionov, S.N. and Fedorov, A.V., 2009. Possible Directions for the Implementation of the Military Policy of the Russian Federation in the Field of International Information Security in Modern Conditions. *Voennaya Mysl'*, 4, 10–15.

Chekinov, S. G. and Bogdanov, S. A., 2015. Прогнозирование характера и содержания войн будущего: проблемы и суждения [Forecasting the Character and Content of Wars of the Future: Problems and Assessments]. *Voennaya Mysl'*, 10, 41–49.

Danyk, Y., Maliarchuk, T. and Briggs, C., 2017. Hybrid War: High-Tech, Information and Cyber Conflicts. *Connections*, 16(2), 5–24. https://doi.org/10.11610/Connections.16.2.01

Fabian, S., 2019. The Russian Hybrid Warfare Strategy – Neither Russian nor Strategy. *Defense & Security Analysis*, 35(3), 308–325. https://doi.org/10.1080/14751 798.2019.1640424

Fabian, S. and Berzins, J., 2021. Striking the Right Balance – How Russian Information Operations in the Baltic States Inform U.S. Strategy in Great Power Competition. *Modern War Institute at West Point*, 4 December. Available from: https://mwi.usma.edu/striking-the-right-balance-how-russian-information-operations-in-the-baltic-states-should-inf orm-us-strategy-in-great-power-competition/

Flook, K., 2009. Russia and the Cyber Threat. *American Enterprise Institute Critical Threats*, 13 May. Available from: www.criticalthreats.org/analysis/russia-and-the-cyber-threat

Galeotti, M., 2019. The Baltic States as Targets and Levers: The Role of the Region in Russian Strategy. *George Marshall European Center for Security Studies: Security Insights*. Available from: www.marshallcenter.org/en/publications/security-insights/bal tic-states-targets-and-levers-role-region-russian-strategy-0

Gerasimov, V., 2016. The Value of Science is in The Foresight: New Challenges Demand Rethinking the Forms and Methods of Carrying Out Combat Operations. *Military Review*, 96(1), 23–29.

Göransson, M. B., 2022. Russian Scholarly Discussions of Nonmilitary Warfare as Securitizing Acts. *Comparative Strategy*, 41(6), 526–542.

Heickerö, R., 2010. *Emerging Cyber Threats and Russian Views on Information Warfare and Information Operations*. Stockholm: Defence Analysis, Swedish Defence Research Agency (FOI).

Klimburg, A., 2011. Mobilising Cyber Power. *Survival*, 53(1), 41–60. https://doi.org/ 10.1080/00396338.2011.555595

Kojala, L. and Keršanskas, V., 2015. The Impact of the Conflict in Ukraine on Lithuanian Security Development. *Lithuanian Annual Strategic Review*, 13(1), 171–189.

Korns, S.W. and Kastenberg, J.E., 2009. *Georgia's Cyber Left Hook*. Carlisle, PA: Army War College Carlisle Barracks Pa Strategic Studies Institute.

Kowalewski, A., 2017. Disinformation and Reflexive Control: The New Cold War. *Georgetown Security Studies Review*, 1. Available from: https://georgetownsecuritystudie sreview.org/2017/02/01/disinformation-and-reflexive-control-the-new-cold-war/

Lane, T., 2014. *Lithuania: Stepping Westward*. New York: Routledge.

Lucas, E., 2016. Cyber in Tallinn. *Center for European Policy Analysis*, 6 June. Available from: http://cepa.org/Cyber-in-Tallinn

Mancuso, J., 2023. *Motives and Tactics, Techniques, and Procedures: Making Sense of the Activities of Russian Advanced Persistent Threat Groups* (Doctoral dissertation, Utica University).

Mattis, J., 2018. *Summary of the 2018 National Defense Strategy of the United States of America*. Washington, DC: United States Department of Defense.

Mattsson, P.A., 2015. Russian Military Thinking – A New Generation of Warfare. *Journal on Baltic Security*, 1(1), 61–70.

Mitchell, L.A., 2012. *The Color Revolutions*. Philadelphia, PA: University of Pennsylvania Press.

Nagorny, A.A. and Shurygin, V.V., 2013. *Defense Reform as an Integral Part of a Security Conception for the Russian Federation: A Systemic and Dynamic Evaluation*. Moscow: Izborsky Club.

Office of Force Transformation, 2005. *The Implementation of Network-Centric Warfare*. Washington, DC: United States Department of Defense.

Piotrowski, M., 2016. The Swedish Counter-Intelligence Report on Hostile Russian Activities in the Region in a Comparative Context. *Polish Institute of International Affairs*, 25(875), 24 March. Available from: www.pism.pl/files/?id_plik=21575

Ploumis, M., 2022. Comprehending and Countering Hybrid Warfare Strategies by Utilizing the Principles of Sun Tzu. *Journal of Balkan and Near Eastern Studies*, 24(2), 344–364.

Shlapak, D.A. and Johnson, M.W., 2016. *Reinforcing Deterrence on NATO's Eastern Flank: Wargaming the Defense of the Baltics*. Santa Monica, CA: RAND Arroyo Center.

United States Department of Defense, 2012. *Joint Publication 3–13 Information Operations*. Washington, DC: United States Department of Defense. Available from: https://irp.fas. org/doddir/dod/jp3_13.pdf

Vasu, N., Ang, B., and Jayakumar, S. eds., 2018. *Drums: Distortions, Rumours, Untruths, Misinformation, and Smears*. London: World Scientific.

4 Russian Cyberspace Operations against Ukraine in the 2022 War

How Effective Have They Been and What Lessons for NATO Can Be Drawn?

Marina Miron and Rod Thornton

Introduction

In what may be seen as the "great cyber competition" between Western states and Russia, one point of contestation has been particularly evident. This was during the Russian invasion of Ukraine, beginning in February 2022. Having the experience of the devastating Russian *NotPetya* attack against Ukraine in 2017 (Chertoff and Kaushik 2023), it was generally expected that Moscow would deploy an almost overwhelming degree of power in the cyberspace arena once the full-scale war had begun. But such an expectation was not borne out. The Russian use of operations in the cyber realm – as far as it can be judged – has been limited in both scale and effectiveness. This chapter, in large part using Russian sources, seeks to understand the reasons for this. It will examine the nature and the degree of success of the cyberattacks that have been conducted against Ukraine and against those NATO states seen to be supporting Kyiv. It will conclude that while, indeed, the Russian use of the power vested in its cyber weapons seems to have been constrained, the threat to the West, more generally from future Russian cyberspace operations, cannot readily be dismissed.

This chapter begins by presenting an understanding of how the Russians conceptualize the notion of warfare in the cyber domain. It then goes on to look at cyber operations against Ukraine in what the Russians call the "cyber-technical" realm, followed by a section on what is known as "cyber-psychological" activities; that is, those designed to influence opinion conducted using IT means. There will then be a discussion of those offensive cyber operations conducted against states supporting Ukraine. Finally, a conclusion will be drawn in terms of making a full assessment of Russia's cyberspace operations in regard to the Ukraine war and what this might portend in terms of future "cyber competition" between Russia and NATO states.

The Russian Understanding of Cyberspace Operations

To fully understand the aims and objectives of Russian offensive cyberspace activity, it must be considered within the broader framework of the general concept of what, in Russian eyes, constitutes "information warfare" or "information

DOI: 10.4324/9781003425304-5

confrontation" (*informatsionnoe protivoporstvo*). The importance of operations in the information domain as a force multiplier for Russian agencies in the areas of security and defense – and in what may be seen as Moscow's current confrontation with "the collective West" (here meaning NATO states and those seen to be aligned with it) – has long been recognized. It is something that has been stressed by both Russian civilian academics and the military for many years.[1]

Much has been written within Russia recently about how best to make use of this force multiplier of information. One of the country's most notable political scientists and contemporary theorists of information war theory, Igor Panarin, has explained that information operations can be divided into both technical and informational aspects. The former considers attacks against IT systems and would be understood in the West as what constitutes "cyberattacks." The Russian term used here to describe such attacks is "cyber-technical." This other aspect, the "cyber-psychological," covers attacks on information that is present on electronic media of all forms (including, and importantly, on social media platforms). In cyber-psychological warfare, information is presented, manipulated, or corrupted in ways that are designed to act against people's opinions and mindsets. Occasionally, both types – cyber-technical and cyber-psychological – are used in coordination to create synergies of effect. According to Panarin, "… sometimes the methods of information and technical influence are carried out in combination with the methods of information and psychological confrontation" (Panarin quoted in Jashibekova 2011). The underlying reason for using these two forms of information warfare (IW) – either together or individually – as another prominent Russian information war theorist, Sergey P. Rastorguev (1999, Ch. 6(1)), has made clear, is that IW "weapons" are highly cost-effective in relation to the advantages, including and most importantly, at the strategic level that they offer to produce.

The Russian military has also highlighted the role that IW can play in its own activities. General Yuri Baluyevsky, who served as the Chief of the Armed Forces General Staff from 2004 to 2008, stressed, for instance, that winning an information war was more important than winning a classical military confrontation. Information wars can remain bloodless and therefore do not invite kinetic retaliation, but they can still produce significant effects against "the principal organs" of the enemy state (RIA Novosti 2017). The current (May 2023) Chief of the General Staff, General Valerii Gerasimov, further elevated the importance of IW as a weapon in a speech he made in 2013. This was summarized in his important article entitled, *The Value of Science in Foresight* (Gerasimov 2013). Similarly, the influential retired senior officers S. G. Chekinov and S. A. Bogdanov (Chekinov and Bogdanov 2010; Bogdanov and Chekinov 2010) have been repeatedly arguing in Russian military publications that the main focus of warfare should be placed on destroying adversary states from *within* using non-kinetic means such as IW instead of trying to achieve such destruction by using kinetic means (which would be difficult, in their eyes, given that Russia's Armed Forces would be up against a more powerful NATO). The importance of information operations has also been repeatedly stressed in other military publications, most notably by Colonel (ret.) Aleksandr Barthosh. Barthosh (2021) underlines the importance of achieving information superiority by influencing an adversary state's information systems and the consciousness of its armed forces personnel and population. This idea builds on those presented earlier by Panarin and Rastorguev and elegantly combines them to broaden the understanding of IW that it should always aspire to encompass the mindset of not only the civilian population but also that of the military. IT means are a vital conduit here. From this perspective, as Margarita Simonyan (2013), the then editor-in-chief of *Russia Today*, put it, "… information weapons are comparable to weapons of mass destruction."

It is important to understand the role of the Russian military in regard to the operationalization of both forms of "cyberattack" – the cyber-technical and the cyber-psychological. It must also be understood that it is the military that is the chief architect of virtually all of Russia's offensive cyberspace activity (with those of the internal security service, the FSB,[2] and the external security service, the SVR,[3] being subordinated to and, to a degree, controlled by those of the military (Lilly and Cheravitch 2020, pp. 140–148). The main protagonist here is the Main Intelligence Directorate of the armed forces, more familiarly known as the GRU.[4] The GRU has experts in the use of both cyber-technical and cyber-psychological attacks (Lilly 2022, pp. 28–29).

The GRU is obviously a finite resource when it comes to the conduct of cyberspace operations against state adversaries. It has a limited number of serving operatives. But what it can do is create critical mass by invoking the assistance of certain cyber adjuncts within Russia which can come under its direction. Here are the likes of "troll farms" or "troll factories" (such as the Internet Research Agency[5]) and a large number of "patriotic hackers" (cyber specialists either press-hanged or cajoled into serving the interests of the GRU). Moreover, the expertise of a large number (over 1,000) of civilian IT firms operating on Russia's cybersecurity market can also be utilized. Thus, the GRU (often supported by the FSB) can bring to bear a considerable number of actors to conduct not just cyberattacks in the technical sense but those that involve both cyber espionage (see Tucker 2023)[6] and information operations aimed at generating influence via IT means. The latter, as the head of the Zecurion analytical center, Vladimir Ulyanov, explains, "include various means of influencing the mood and behavior of the population of countries" (Ulyanov quoted in Kolomychenko 2017). Ulyanov recognizes, as has the Russian military, that the more developed the country, the more vulnerable it is to cyberattacks, given that with the development of any state comes a dependency on IT means connected to the Internet. As development increases, the available target set for the GRU becomes larger (Ulyanov quoted in Kolomychenko 2017).

The Russian military is one that emphasizes the importance of IW to a degree that Western militaries do not. The former sees it as a core non-kinetic weapon in its arsenal, while the latter see it as merely peripheral to its core kinetic activities. One example of just how seriously the Russian military takes the whole aspect of cyber-related operations comes with the fact that in 2015, Russian Defense Minister Sergei Shoigu announced the creation of a self-sufficient body of information operations troops in (Russian-annexed) Crimea in the fall of 2015. These troops were set up to pursue national interests in the information sphere. In the end, over 100 formations, units, and organizations came to be part of or associated with these Crimean IW troops (TASS 2015).

Despite, though, all the emphasis placed on the conduct of IW, it is being made clear by Russian sources that improvements can still be made. This is now specifically based on experience gained in Moscow's interventions in Ukraine. Following the initial annexation of Crimea in 2014, there were the first signs of Russia losing the information war against the West due to a perceived inability to adapt and improve its operations in both technological and media-content terms (Snegovaya 2015). In 2017, Panarin (2017) stressed that Russia was still lagging in the cyber-psychological domain, calling for a need to learn from what the United States was doing; particularly in terms of setting up a Russian equivalent of the U.S. Cyber Command. This would make it easier to master the interplay of both technical and psychological elements across a number of Russian cyber agencies (Panarin 2017).

While it was felt that something had to be done, it is not immediately clear, however, just what these changes might have been. But certainly, some notable successes appeared to have taken place between 2015 and 2018. In this period, Russian offensive cyber groups and/or their affiliates carried out some major cyber operations (e.g., hacking the U.S. national elections of 2016, which included both cyber-technical and cyber-psychological elements (see Yourish et al. 2018; Yang 2019, pp. 103–113). Some changes to the structures managing Moscow's offensive cyber activities were, however, more obvious. In 2018, Shoigu announced the creation of a new formation within the military – the Russian Information Security Troops (Osipov 2017). This is supposedly designed to resemble the 7th Directorate of the General Staff of Soviet times, which was responsible for propaganda. However, unlike the old 7th Directorate, which was subordinate to the Main Political Directorate of the Soviet Armed Forces, these new "troops" are subordinate to the Main Operational Directorate of the General Staff. It seems that the GRU has been brought into this new formation. It is a structure that appears to unite several units into one. Previously, various GRU units had responsibility for information operations without a clear overarching direction from a single body. Included, of course, in this body are those troops tasked with conducting tasks in cyberspace (Osipov 2017).

A History of Russian Cyberattacks against Ukraine

When the 2022 invasion began – the "special military operation" – a major use of Russian IW capabilities was expected. Ukraine had certainly been subject to a series of Russian cyber-technical attacks in the years prior to 2022 (and in addition to the aforementioned and very high-profile *NotPetya* assault). For instance, the Computer Emergency Response Team of Ukraine (CERT-UA 2022a) had reported several hacking attempts, including the use of a fraudulent online tax-collection document that notified users of unpaid taxes (*povidomlennya pro nesplatu podatky*). This led the unwary to install Cobalt Strike Beacon, the capabilities of which ranged from command execution to file transfer. This attack was sourced to the group known as UAC-0098 (see Rahman 2021). While UAC-0098 was not directly linked to any

Russian hacker group, it has been known to have acted as an initial access broker to a Russian hacker group known as FIN12/WIZARD SPIDER. This same group had also impersonated different legitimate entities such as the Ukrainian cyber police and even Microsoft (Google Threat Analysis Group 2022).

Moscow's targeting of critical national infrastructure (CNI), such as Ukraine's power grid, had previously been observed in 2015 and 2016. Then *BlackEnergy3* malware was used to exploit Microsoft Word's macro-feature giving the hackers access to IT systems responsible for distributing electrical power (Zetter 2016; BBC 2017). Such attacks showed a high degree of sophistication due to their coordinated nature and the consideration of all possible mitigation plans by the target (Zetter 2016). The Ukrainian power grid was also later targeted by the Russian hacker group, *Sandworm*, controlled by the GRU, and specifically this agency's Unit 74455. This attack employed a modified piece of *Industroyer* malware (Greenberg 2020, 2022a).

In the cyber-psychological sphere, other work had been done by ATP28, also known as *FancyBear.* This GRU hacker group has been linked to numerous high-profile attacks on foreign political entities in the United States, Germany, and various other European states (FireEye 2017). In Ukraine, ATP28's activities included defacing social media profiles and government websites, including during Kyiv's presidential elections in 2014 (Google Threat Analysis Group 2022).

Offensive Cyberspace Operations against Ukraine in the Cyber-Technical Realm

Given this "rich" history of Russian cyber operations against Ukraine, it was generally expected that once the war of 2022 started, such attacks and, indeed, their more refined brethren would be similarly unleashed against Ukraine. But this did not seem to occur. Of course, immediately after the ground invasion, there was the high-profile hack of the ViaSat KA-SAT satellite network serving Ukraine, which targeted its ground terminals. This left not only Ukrainians but also many consumers across wider Europe, who were reliant on the same system, without an Internet connection. Additionally, official Ukrainian websites suffered from severe DDoS attacks and became inaccessible (Burgess 2022; Urban 2022). Ukraine's military communications were also targeted, not least due to the ViaSat hack (Zhora 2022; Burgess 2022). Although no official attribution could occur for these attacks and others, many U.S. and Ukrainian officials pointed the finger at the obvious culprit, Russia. But as a senior official at Ukraine's cybersecurity agency, Viktor Zhora, suggested, it was difficult to prove that Russia had, indeed, sponsored the attacks. It was plausible, though, to believe that active persistent threat (APT) groups, most likely directed by the GRU, were behind the attack (Zhora 2022).

Many more cyberattacks took place but without the high-profile effects of the ViaSat and the DDoS operations against official sites. A Microsoft report, for instance, disclosed that a veritable "cyber storm" had been conducted to support the initial Russian offensive (Satter et al. 2022). One element here, and noted on several occasions, was the coordination of cyberattacks with kinetic military strikes, most notably when Russian missiles struck Kyiv's TV tower on 1 March

2022, simultaneously as a cyberattack against the TV station's IT systems. The same coordinated activity was seen with several further cyberattacks and kinetic strikes on the infrastructure of Ukrainian media companies.

Other more low-profile attacks were officially noted, often against Ukraine's outdated IT infrastructure, such as targeting the routers used in the state transportation system (Greenberg 2022c). There was, for example, an effective phishing attack against the Ukrainian Railway Transport Organization that was attributed to an entity known as UAC-0140 (CERT-UA 2022b, Antonyuk 2022).

Considerable cyber espionage activity was also observed. These showed how Russia's cyberspace forces were attempting to integrate their activities across several vectors and with some degree of success (Satter et al. 2022). Beyond the general, moreover, some quite specific cyber espionage activities were noted. For instance, a new malware toolkit, *Pipedream*, was uncovered (though not officially attributed to Russia). This enables the end user of such malware to target critical infrastructure in "practically any industrial environment, from manufacturing to water treatment" (Greenberg 2022b).

But while most cyber-technical attacks were themselves in the category of low-profile, there was this aspect of attempting to produce a force-multiplier effect by accompanying them with linked kinetic attacks. According to Brigadier General Yuri Shchyhol, the head of Ukraine's State Special Communication and Information Protection Service (*Derzhspetszviazok*), about one-fifth of all alleged Russian cyberattacks were aligned to kinetic strikes – particularly against CNI (Shchyhol quoted in Anotniyk 2022). Moreover, there has been evidence of not just coordination between Russian cyber actors and the regular armed forces but also between the different cyber actors themselves. This makes actual attribution difficult. But even if digital forensic evidence is lacking, the pool of potential perpetrators is fairly small. Most, though, have been clearly identified as linked to the GRU (see Sherman 2022).

Offensive Cyberspace Operations against Ukraine in the Cyber-Psychological Realm

Along with attacks in the technical arena, other Russian cyber tools have been used during the invasion to impact the psyche of the Ukrainian population in the service of Moscow's interests. Cyberspace operations have been used to spread disinformation and to skew the public perception of specific political and military figures in Ukraine. Deepfakes have notably been employed in this regard. Deepfakes can be seen as a cross-over of cyber-technical and cyber-psychological means. The Russian attempts at generating such deepfakes, however, have been clumsy. For instance, on 2 March 2022, Ukraine's Center for Strategic Communication warned of the possibility that a deepfake of President Volodymyr Zelensky would appear on media outlets. Indeed, such a video did then appear shortly after on both Facebook and YouTube. In the video, "Zelensky" was seen calling on Ukrainian troops to lay down their arms. Shortly after its appearance, the "real" Zelensky had to post a Facebook video denying the claims "he" had made in the video (Simonite 2022). What is perhaps worrisome here is that neither Facebook nor YouTube had then the necessary algorithms for detecting such synthetic videos – although solutions do exist and were duly utilized post hoc (Nguyen et al. 2022). It should be expected, though, and as generative AI means continue to improve, that the quality of deepfakes will also improve (Hay Newman 2023). Russian cyber actors will thus doubtless be able to enhance their attempts with the use of deepfakes in the future.

Further, Russian operations in the cyber-psychological space made use of troll farms to manipulate public opinion on a large scale in Ukraine (The Moscow Times 2023). The likes of the Internet Research Agency have been exploiting social media platforms such as Telegram, Twitter/X, Facebook, and TikTok to spread content related to supporting the Kremlin's views as well as calling on sympathizers in Ukraine (and beyond) to target the Kremlin's opponents on social media (UK Government 2022). Such troll farms, however, were not just operating on Russian territory. One, for example, was discovered by Ukrainian authorities in the Ternopil'ska Oblast'. A resident there had created a troll farm to spread damaging (pro-Russian) content (Main Directorate of National Policy of Ukraine 2023).

Beyond the output of the more organized troll farms during the war, the Russians also made use of individual "patriotic hackers" to spread disinformation. While there has been an issue with aligning the activities of such hackers with the strategic messaging required by Moscow, they have, it seems, provided a degree of effect. A hacker from [Russian occupied] Donetsk, for instance, known as Joker, hacked Valerii Zaluzhny's (Commander in Chief of the Ukrainian Armed Forces) Instagram account and posted transcripts of Zaluzhny's conversations with several women on his Telegram channel (News.ru 2022; Kaverin 2022). It is assumed that this was done in an attempt to discredit Zaluzhny and to damage his reputation, most notably in the eyes of Western audiences, by depicting him as someone immoral. In contemporary parlance, this is referred to as "doxing" (News.ru 2022).

Offensive Cyberspace Operations against States Supporting Ukraine

In looking at the breadth of Russian cyberspace operations against Ukraine in the 2022 war, it must also be borne in mind that the targets can also be beyond Ukraine's borders. The scope of these operations has been very wide, and this chapter can only refer to some of them. They can be seen, though, as very much part of the Russian strategic logic. Cyber actors based within Russia have been noted as targeting – using both cyber-technical and cyber-psychological methods – countries that support Ukraine, most notably the United States and several Eastern European countries. These attacks, at least the publicly acknowledged ones, may be seen, though, as being merely in the "nuisance" category. One example was observed in what became known as the Russian-sourced KillNet attack. This hacked the U.S. FBI's database and stole the personal details, including the Medical IDs and social media passwords, of some 10,000 federal agents (McNulty 2022). A few days later, the Russian hacker group RaHDit (aka "Evil Russian Hackers") published a list of over 100 people in the West involved in sensitive activity related to the Ukraine war, including NATO military personnel attached to its cyber force. One of the RaHDit group members had blamed such NATO personnel for being behind cyberattacks on Russia, which had been re-routed through Ukrainian sources (Shakirov 2022). The ability of Russian cyber actors to create such breaches of security show how NATO personnel involved in cyber operations can expose themselves to risk, even to their physical security.

Other types of cyberspace activity against NATO states have included attacks on the IT systems of Europe's oil and gas sector. While these appear to have been conducted by ransomware groups, there are possible links to the Russian internal security service, the FSB, which has been known to exploit cybercriminal gangs. These attacks, however, did not meet the magnitude expected in prior warnings by U.S. and NATO officials. More than anything, they created more "background noise" than damage compared to the likes of *NotPetya* (Martin 2022).

Of course, officially attributing the activities of some of these hackers targeting NATO and individual NATO states back to Russia is difficult. But it is not hard to imagine that they might be the work of an overall Russian "cyber force." Certainly, some are believed to have links to the GRU. One such is the Ghostwriter group, which operates from Belarus and has engaged in disinformation activities targeted at Poland and its support for Ukraine. This group tried to hack the email addresses and social media accounts of public figures supportive of Ukraine. This was done in presumed attempts to discredit them publicly or to seek information that could be used to build a degree of *kompromat* material to be used later (Sebbagh 2023).

Assessing Russia's Cyberspace Operations in the Ukraine War

Overall, in looking at the various cyberattacks conducted during the war by or linked to Russian cyber actors, the general conclusion has been that they have not been as damaging as might be expected. If it is understood just how much emphasis the Russian military has been giving to operations in cyberspace and just how many cyberattacks were conducted against Ukraine before the 2022 war, why were so few and of any consequence once the war began? This question is especially pertinent when it comes to cyber-technical attacks. The view of many experts that a major cyberwar against Ukraine and against those countries supporting it would come to pass has just not been borne out (Browne 2022).

One reason for this may be that the number of attacks and the damage caused by them is simply not known. For security or public interest reasons, Kyiv may not want to officially disclose what Russian cyber offensives have accomplished. It will be important for the Ukrainian authorities to hide the effects, not just for propaganda purposes but also to deny the Russians the ability to conduct "battle damage assessments" (BDA) – that is, to know the results of their cyber operations. Denying BDA would make it harder for the Russian side to judge what operations to continue with and which to give up on, as effort is simply being wasted. As such, offensive cyber "success" for the Russians could not be reinforced or failure discontinued.

There is a more likely reason, though, for the limited effect of Russian cyberspace operations. This is that, in light of the experience gained from years of being subject to Russian cyberattacks, the Ukrainian authorities had actually developed very good defenses for both its military and civilian-sector IT systems. Indeed, the cyber agencies of Western states had also been providing assistance both before the 2022 war and once it commenced. The likes of Microsoft (Knutson 2022) and the U.S. government had deployed their own cyber defense teams to Ukraine (Siddiqui 2023).

Another factor involved here might be the diminishing quality of Russian cyberattacks. The GRU's offensive cyber operations have not been as sophisticated as those of years past – such as with *NotPetya*. Given that the transitory nature of offensive cyber tools means that "a constant (re)investment is required for the developer of sustainable, constant offensive capability" (Smeets 2018, p. 32), the GRU may have become lax; neither updating its own capabilities nor encouraging/incentivizing its adjunct cyber players to hone theirs. Indeed, it seems that rather than relying on sophisticated pieces of malware, which take both time and resources to develop, the Russian aim in Ukraine has been to simply overwhelm the adversary's cyber defenses with sheer mass. Moreover, the GRU appears to have utilized "out-of-the-box" and quickly deployable solutions to specific problems to increase the pace of its operations targeting Ukrainian IT systems. In such cases, unsophisticated wipers (designed to erase data from devices, making them unusable) were used *en masse* to achieve an immediate effect (Greenberg 2022c). This suggests a departure from the cyber modus operandi apparent against Ukraine prior to 2022: nuanced sophistication appears to have been replaced by cyber brute force.

Another possible reason for the limited cyber assault on Ukraine relates to the fact that the use of cyber actors has been so broad. The Kremlin has used not just the GRU but also troll farms, patriotic hackers, cybercriminals, and general cyber-sympathizers aligned with the Russian cause. There has thus been, at best, a loosely connected network of different participants. However, while creating a welcome degree of mass, the lack of a structure to control and coordinate the range of actors involved may be seen to have hampered the properly coordinated execution of operations (Lewis 2022).

A final reason why the expected cyber onslaught against Ukraine has not materialized may be down to the fact that Russian cyber actors are keeping their "powder dry." That is, if they do have truly sophisticated malware at their disposal, they may not want to deploy it against Ukrainian targets because, once used, counters can then be developed. Instead, such malware would be saved for any possible future conflict with NATO. In such a conflict, the Russian military would want to create a cyber "shock and awe" effect that might generate a devastating – and perhaps "war-winning" – effect (Kofman et al. 2022, p. 5). This cannot happen if Russia has already shown its "cyber hand" in the war with Ukraine.

In terms of cyber-psychological operations undertaken post-invasion, these may be seen as less than successful. The very fact that Russia has gone to war for its supposed interests vis-à-vis Ukraine will have undoubtedly undermined its ability to change the minds of any waverers either within Ukraine or among the Western states supporting it. Moreover, the simple blocking of Russian media websites such as Russia Today and Sputnik (Deutsch 2022) has gone a long way to effectively curtailing Russia's ability to exert influence on Western audiences.

The Future

In looking at the Russian cyberattacks against Ukraine and its NATO state supporters as part of the 2022 invasion, certain characteristics have been evident. The range of attacks has been notably broad. This runs in line with the understood approaches of the Russian military: they are seen to work best as an amalgam of actions in both the technical and psychological realms. Some attacks have been aligned with distinct military objectives, while others have sought to create a psychological effect. Included here were attempts to discredit public figures and spread disinformation. Despite warnings, though, that Ukraine might be subject to some sort of cyber-Armageddon once the war began, the reality appears to have been different.

But this does not mean that Ukraine and NATO, more broadly, should drop their cyber guard where Russia is concerned. There is a – perhaps well-substantiated – concern that a prolonged war in Ukraine could make Russia more aggressive in the cyber domain. Capabilities might be honed in light of experience gained by the Russian side, and investment in cyber potential may increase as Moscow reels from setbacks on the battlefield and what it sees as provocative actions by NATO states. U.S. officials have warned, indeed, that even the U.S. homeland might be threatened in future by Russia's offensive cyber operations (Demarest 2022b).

Russia remains a cyber threat to NATO and its core states. In March 2023, General Paul Nakasone, commander of U.S. Cyber Command and director of the National Security Agency, noted that Russia remained a "very capable [cyber] adversary" (Nakasone quoted in Demarest 2023). And, according to Max Smeets (2022, pp. 51–54), Russia still belongs to the category of the most "dangerous" states in terms of cyber capabilities. This is said to be due to the relatively large organizational and financial resources Moscow devotes to cyberspace operations. Additionally, the military will always, of course, be able to leverage the assets of a large and mature Russian private IT sector which will have the capabilities to develop sophisticated malware. The military, moreover, will always have a rich pool of people from which to "recruit" cyber specialists. Observing activity at the Positive Hack Day cybersecurity forum – the Russian equivalent of the annual DEF CON cyber convention in the United States – held in Moscow in May 2022 is perhaps instructive. This forum serves not only as a means of demonstrating Russia's adaptability to technological isolationism (imposed by the sanctions regime), but it also gives the intelligence services a unique opportunity to recruit from Russia's leading IT enterprises and universities (see Sherman 2023). Perhaps more importantly, Positive Hack Day fulfills a more nuanced role by bringing together Russia's cyber potential to determine ways forward amid the current conditions. For external observers, it represents a show of "cyber force," signifying that Russia still has the capacity to inflict a significant degree of effect in its offensive cyberspace operations. It must also be borne in mind, however, that such capacities in the field of Russian IW will undoubtedly have been badly affected by the substantial brain-drain that followed Russia's 2022 invasion of Ukraine and which had already begun in modest order after that in 2014 (see Pinna 2022). Only some three weeks after the second invasion, about 100,000 IT specialists (or 10% of their number in Russia) had left the country, according to Russia's minister for digital development, Maksut Shadayev (quoted in Ilyushina 2023).

Thus, when looking at where the land lies in relation to Russian cyberspace capabilities that could one day be employed in force and *en masse* against Western states, the jury might be said to be out. Such capabilities may be a "known unknown." And alongside capabilities, of course, has to run intent: just how much damage does Moscow want to create in NATO states with its offensive cyber operations? This is also a "known unknown." But the Russians do understand just what a force multiplier the cyber tool represents, particularly as AI comes to play a greater role in enhancing the capabilities of this tool (Thornton and Miron 2020). Both NATO states and, indeed, the Alliance's partner states – and with good reason – remain on high alert to the possibility of future highly effective Russian cyberattacks in both the technical and psychological realms (Paul 2022; Politico 2022; Demarest 2022a).

Notes

1 The idea of "information war" was much discussed in Soviet times. For instance, Evgenii Messner's *Myatezhvoyna* ("Rebellion War") was arguably the first concept focusing on influencing the consciousness of an adversary state's population. This chapter, however, focuses only on contemporary theorists who take into account the cyber dimension of such "information war." For more on "rebellion war," see E. Messner (1960) and S. Anchukov (2000).
2 *Federal'naya Sluzhba Bezopasnosti* or Federal Security Service is the main successor of the Soviet KGB. Its operations take place within Russia and internationally.
3 *Sluzhba Vneshnei Razvedki* or Foreign Intelligence Service is tasked with intelligence activities outside Russia.
4 *Glavnoye Upravleniya* (GU) or Main Directorate – formerly known as GRU – is Russia's military intelligence arm. The authors will use the acronym "GRU" given its continued use in English-language publications.
5 The Internet Research Agency was founded by the late Evgeni Prigozhin, the chief of the Wagner private military company. It has been used extensively to spread disinformation and misinformation across the globe in the service of Russian strategic interests (The Moscow Times 2023).
6 The United States recently uncovered a major cyberespionage operation in the United States and NATO led by the FSB (Tucker 2023).

References

Anchukov, S. V., 2000. *Sekrety Myatezhnoi Voyny. Rossia na Rubezhe Stoletii* [*The Secrets of Rebellion War. Russia at the Turn of the Century*]. Moscow: Veonizdat.
Antonyuk, D., 2022. Ukrainian Railway, State Agencies Allegedly Targeted by DolphinCape Malware. *The Record*, 12 December. Available from: https://therecord.media/ukrainian-railway-state-agencies-allegedly-targeted-by-dolphincape-malware/

Barthosh, A. A., 2021. Gibridnaya, Skrytnaya, Nepredskazuyemaya [Hybrid, Covert, Unpredictable]. *Nezavisimoye Voennoye Obozreniye*, 12 August. https://nvo.ng.ru/gpolit/2021-08-12/1_10_11_1153_hybrid.html

BBC, 2017. Ukraine Power Cut "Was Cyber-Attack." *BBC*, 11 January. Available from: www.bbc.com/news/technology-38573074

Bogdanov, S. A. and Chekinov, S. G., 2010. Asimmetrichnye Deistviya Po Obespecheniyu Voennoi Bezopasnosti Rossii [Asymmetric Actions for Ensuring Russia's Military Security]. *Voennaya Mysl'*, 3, 13–22.

Browne, R., 2022. The World is Bracing for a Global Cyberwar as Russia Invades Ukraine. *CNBC*, 25 February. Available from: www.cnbc.com/2022/02/25/will-the-russia-ukraine-crisis-lead-to-a-global-cyber-war.html

Burgess, M., 2022. A Mysterious Satellite Hack Has Victims Far Beyond Ukraine. *Wired*, 23 March. Available from: www.wired.com/story/viasat-internet-hack-ukraine-russia/

Chekinov, S. G. and Bogdanov, S. A., 2010. Vliyaniye Asimmetricheskikh Deystviy Na Sovremennuyu Bezopasnost' Rossii [The Influence of Asymmetric Actions on Russia's Contemporary Security]. *Vestnik Akademii Voennykh Nauk*, 1, 46–53.

Chertoff, M. and Kaushik, A., 2023. The Unheralded Success Story of Ukraine's Cyber-Defenses. *EU Observer*, 1 March. Available from: https://euobserver.com/opinion/156766

Computer Emergency Response Team of Ukraine, 2022a. Uvaga! Novi Kiberataky Cherez Rozsylannya Nebezpechnykh Failiv: Slovmysnyky Vykorystovyyut' Temy Shtrafiv ta Yadernogo Terorizmy [Attention! New Cyberattacks through Dissemination of Dangerous Files: Criminals Exploit Issues Related to Tax Penalties and Nuclear Terrorism]. *State Services of Special Communications and Information Protection of Ukraine*, 21 June. Available from: https://cip.gov.ua/ua/news/uvaga-novi-kiberataki-cherez-rozsilannya-nebezpechnikh-failiv-zlovmisniki-vikoristovuyut-temi-shtrafiv-vid-podatkovoyi-ta-yadernogo-terorizmu

Computer Emergency Response Team of Ukraine, 2022b. Kiberataka Na Derzhavni Organizatsii z Vykorystannyam Tematyky Iran'skykh Droniv-Kamikadze Shahed-136 ta Shkidlyvoyi Programy DolphinCape (CERT-UA#5683) [Cyber-attack on Government Organizations Using the Theme of Iranian Shahed-136 Kamikaze Drones and DolphinCape Malware (CERT-UA#5683)]. *State Services of Special Communications and Information Protection of Ukraine*, 8 December. Available from: https://cert.gov.ua/article/3192088

Demarest, C., 2022a. Feared Russian Cyberattacks Against US Have to Yet Materialize. *C4ISRNET*, 29 April. Available from: www.c4isrnet.com/cyber/2022/04/29/feared-russian-cyberattacks-against-us-have-yet-to-materialize/

Demarest, C., 2022b. Prolonged War May Make Russia More Cyber Aggressive, US Officials Say. *C4ISRNET*, 17 June. Available from: www.c4isrnet.com/cyber/2022/06/17/prolonged-war-may-make-russia-more-cyber-aggressive-us-official-says/

Demarest, C., 2023. Russia Remains a "Very Capable" Cyber Adversary, Nakasone Says. *C4ISRNET*, 7 March. Available from: www.c4isrnet.com/cyber/2023/03/07/russia-remains-a-very-capable-cyber-adversary-nakasone-says/

Deutsch, J., 2022. RT, Sputnik Content Officially Banned Across European Union. *Bloomberg*, 2 March. Available from: www.bloomberg.com/news/articles/2022-03-02/rt-sputnik-content-officially-banned-across-european-union

FireEye iSight Intelligence, 2017. *ATP28 at The Center of the Storm – Special Report*. Mandiant, January. Available from: www.mandiant.com/sites/default/files/2021-09/APT28-Center-of-Storm-2017.pdf

Gerasimov, V., 2013. Tsennost' Nauki v Predvidenii [The Value of Science in Foresight]. *Voenno-Promyshlennyi Kur'er*, 8(476). Available from: https://vpk.name/news/85159_cennost_nauki_v_predvidenii.html

Google Threat Analysis Group, 2022. Initial Access Broker Repurposing Techniques in Targeted Attacks Against Ukraine. *Google*, 7 September. Available from: https://blog.goo gle/threat-analysis-group/initial-access-broker-repurposing-techniques-in-targeted-atta cks-against-ukraine/

Greenberg, A., 2020. *Sandworm: A New Era of Cyberwar and the Hunt for the Kremlin's Most Dangerous Hackers*. New York: Doubleday.

Greenberg, A., 2022a. Russia's Sandworm Hackers Attempted the Third Blackout in Ukraine. *Wired*, 16 April. Available from: www.wired.com/story/sandworm-russia-ukra ine-blackout-gru/

Greenberg, A., 2022b. Feds Uncover a "Swiss Army Knife" For Hacking Industrial Control Systems. *Wired*, 27 April. Available from: https://wired.com/story/pipedream-ics-malware/

Greenberg, A., 2022c. Russia's New Cyberwarfare in Ukraine is Fast, Dirty and Relentless. *Wired*, 10 November. Available from: www.wired.com/story/russia-ukraine-cyberattacks-mandiant/

Hay Newman, L., 2023. NSA Cybersecurity Director Says "Buckle Up" for Generative AI. *Wired*, 28 April. Available from: www.wired.com/story/nsa-rob-joyce-chatgpt-security/

Ilyushina, M., 2023. Russia Eyes Pressure Tactics to Lure Fleeing Tech Workers Home. *The Washington Post*, 8 March. Available from: www.washingtonpost.com/world/2023/03/ 08/russia-employers-intimidation-workers-war/

Jashibekova, J., 2011. Igor' Panarin: V Informatsionnykh Voynakh u Rossii Dolzhen Byt' Krepkii Shchit IO [Igor Panarin: In Information Wars Russia Must Have a Strong Shield of IO]. *Argumenty I Fakty*, 24 March. Available from: https://aif.ru/society/igor_panarin_ v_informacionnyh_voynah_u_rossii_dolzhen_byt_krepkiy_schit_i_o

Kaverin, D., 2022. Donetskii Haker "Dzhoker" Pokazal Perepiski Glavkoma Glavkoma VSU Zaluzhnogo s Devushkami [A Hacker "Joker" from Donetsk Showed Correspondence of the Commander-in-Chief of the Ukrainian Armed Forces Zaluzhnyi with Ladies]. *Gazeta.ru*, 5 November. Available from: www.gazeta.ru/social/news/2022/11/05/18966 139.shtml

Knutson, J., 2022. Microsoft: Russia Has Conducted Hundreds of Cyberattacks Against Ukraine. *AXIOS*, 27 April. Available from: www.axios.com/microsoft-russia-ukraine-invasion-cyber-12c8aa49-e1d6-40f9-aa13-962baa935acd.html

Kofman, M., Connolly, R., Edmonds, J., Kendall-Taylor, A., and Bendett, S., 2022. Assessing Russian State Capacity to Develop and Deploy Advanced Military Technology. *Center for a New American Security*, October. Available from: www.cnas.org/publications/reports/ assessing-russian-state-capacity-to-develop-and-deploy-advanced-military-technology

Kolomychenko, M., 2017. V Internet Vveli Kibervoiska [Cyber Troops Have Entered the Internet]. *Kommersant*, 10 October. Available from: www.kommersant.ru/doc/ 3187320

Lewis, J. A., 2022. Cyber War and Ukraine. *Center for Strategic and International Studies*, 16 June. Available from: www.csis.org/analysis/cyber-war-and-ukraine

Lilly, B., 2022. *Russian Information Warfare*. Annapolis, MD: Naval Institute Press.

Lilly, B. and Cheravitch, J., 2020. The Past, Present and Future of Russia's Cyber Strategy and Forces. *In*: T. Jancarkova, L. Lindstrom, M. Signoretti and I. Tolga, eds. *20/20 Vision: The Next Decade*. Tallinn: NATO, 129–155. Available from: https://ccdcoe.org/ uploads/2020/05/CyCon_2020_8_Lilly_Cheravitch.pdf

Main Directorate of the National Police of Ukraine, 2023. Kiberpolitseys'ki Spil'no z Operatyvnykamy SBU Vykryly Botofermu Cherez Yaku Poshyryuvaly Vorozhyy Kontent [Cyber Police Together with SBU Operatives Uncovered a Troll Farm Through Which Hostile Content was Distributed]. *National Police of Ukraine*, 3 January. Available from: https://tp.npu.gov.ua/news/kiberpolitseiski-spilno-z-operatyvnykamy-sbu-vykryly-botofermu-cherez-iaku-poshyriuvaly-vorozhyi-kontent

Martin, A., 2022. Russian Hackers Targeted Petroleum Refining Company in NATO State. *The Record*, 20 December. Available from: https://therecord.media/russian-hackers-targe ted-petroleum-refining-company-in-nato-state/

McNulty, T., 2022. Russian Hackers Claim to Have Infiltrated FBI with Names and Bank Details Exposed. *Daily Express*, 16 December. Available from: www.express.co.uk/news/ world/1710122/russian-hackers-fbi-infiltrate-data-killnet

Messner, E, 1960. *Myatezh – Imya Tretei Mirovoi Voyny* [Rebellion – The Name of the Third World War]. Moscow: Moscow Publishing House.

News.ru, 2022. Hakery Vzlomali Instagram Glavkoma VSU Zaluzhnogo [Hackers hacked Instagram of Commander-in-Chief of the Armed Forces of Ukraine Zaluzhny]. *News.ru*, 1 December. Available from: https://news.ru/europe/sholc-nazval-tri-prichiny-ne-postavl yat-tanki-na-ukrainu/

Nguyen, T. T., Hung Nguyen, Q. V., Tien Nguyen, D., Thanh Nguyen, D., Huynh-The, T., Nahavandi, S., Tam Nguyen, T., Pham, Q. V., and Nguyen, C. M., 2022. Deep Learning for Deepfakes Creation and Detection: A Survey. *Computer Vision and Image Understanding*, 223, 1–14. https://doi.org/10.1016/j.cviu.2022.103525

Osipov, S., 2017. Tsifrovoi Spetsnaz Shoygu. Chemy Protivostoyat Voyska Informatsionnykh Operatsiy? [Shoygu's Digital Special Forces. To What Will the Information Troops Offer Opposition?]. *Argumenty I Fakty*, 2 March. Available from: https://aif.ru/society/army/ cifrovoy_specnaz_shoygu_chemu_protivostoyat_voyska_informacionnyh_operaciy

Panarin, I., 2017. Kak Vesti Voynu Novostei [How to Wage a News War]. *Argumenty I Fakty*, 3 February. Available from: https://aif.ru/politics/opinion/kak_vesti_voynu_n ovostey?from_inject=1

Paul, K., 2022. Russia's Slow Cyberwar in Ukraine Begins to Escalate, Experts Say. *The Guardian*, 2 April. Available from: www.theguardian.com/world/2022/apr/01/russia-ukra ine-cyberwar

Pinna, M., 2022. Russia's Brain Drain: War with Ukraine Prompts Tens of Thousands to Flee Abroad. *Euronews*, 10 June. Available from: www.euronews.com/2022/06/10/rus sia-s-brain-drain-thousands-flee-abroad-since-start-of-war-with-ukraine

Politico, 2022. The World Holds Its Breath for Putin's Cyberwar. *Politico*, 23 March. Available from: www.politico.com/news/2022/03/23/russia-ukraine-cyberwar-putin-00019440

Rahman, A., 2021. Defining Cobalt Strike Components So That You Can BEA-CONfident in Your Analysis. *Mandiant*, 21 October. Available from: www.mandiant.com/resources/ blog/defining-cobalt-strike-components

Rastorguev, S. P., 1999. *Informatsionnaya Voyna: Problemy I Modeli* [*Information War: Problems and Models*]. Moscow: Radio and Communication.

RIA Novosti, 2017. Baluyevsky: Pobeda v Informatsionnoi Voine Vazhnee, Chem v Klassicheskoi [Baluyevsky: Victory in Information War is More Important than a Classical One]. *RIA Novosti*, 22 February. Available from: https://ria.ru/20170222/148 8611839.html

Satter, R., Bing, C. and Pearson, J., 2022. Microsoft Discloses Onslaught of Russian Cyber Attacks on Ukraine. *Reuters*, 27 April. Available from: https://reuters.com/technology/ microsoft-discloses-onslaught-russian-cyberattacks-ukraine-2022-04-27/

Sebbagh, D., 2023. Cyber-attacks Have Tripled Last Year, Says Ukraine's Cyber Security Agency. *The Guardian*, 19 January. Available from: www.theguardian.com/world/2023/ jan/19/cyber-attacks-have-tripled-in-past-year-says-ukraine-cybersecurity-agency

Shakirov, E., 2022. Hakery RaHDit Opublikovali Dannye Ofitserov iz "Kibervoisk NATO" [RaHDiT Hackers Published the Data of "NATO Cyber Command" Officers]. *Gazeta*, 29 December. Available from: www.gazeta.ru/tech/news/2022/12/29/19383 319.shtml

Sherman, J., 2022. GRU 26165: The Russian Cyber Unit that Hacks Targets On-Sight. *Atlantic Council*, 18 November. Available from: www.atlanticcouncil.org/content-series/ tech-at-the-leading-edge/the-russian-cyber-unit-that-hacks-targets-on-site/

Sherman, J., 2023. Russia's Largest Hacking Conference Reflects Isolated Cyber Ecosystem. *Brookings*, 12 January. Available from: www.brookings.edu/techstream/russias-largest-hacking-conference-reflects-isolated-cyber-ecosystem/

Siddiqui, Z., 2023. U.S. Deploys More Cyber Forces Abroad to Help Fight Hackers. *Reuters*, 25 April. Available from: www.reuters.com/technology/us-deploys-more-cyber-forces-abroad-help-fight-hackers-2023-04-25/

Simonite, T., 2022. A Zelensky Deepfake Was Quickly Defeated. The Next One Might Not Be. *Wired*, 17 March. Available from: www.wired.com/story/zelensky-deepfake-faceb ook-twitter-playbook/

Simonyan, M., 2013. Simonyan: Informatsiya kak Oruzhye Ispol'zuyetsya Temi, Kto Imeet Vozmozhnost' [Simonyan: Information as a Weapon Used by Those Who Have the Possibility]. *Rossiyskaya Gazeta*, 3 July. Available from: https://rg.ru/2013/07/03/simon ian.html

Smeets, M., 2018. A Matter of Time: On the Transitory Nature of Cyberweapons. *Journal of Strategic Studies*, 41(1–2), 6–32. https://doi.org/10.1080/01402390.2017.1288107

Smeets, M., 2022. *No Shortcuts: Why States Struggle to Develop a Military Cyber-Force*. London: C. Hurst & Co. (Publishers) Ltd.

Snegovaya, M., 2015. Pochemu Kreml' Proigryvaet Informatsionnyyu Voyny na Zapade [Why Russia is Losing in Information War in the West]. *Vedomosti*, 9 September. Available from: www.vedomosti.ru/opinion/articles/2015/09/09/608046-pochemu-kreml-proigriv aet-informatsionnuyu-voinu-zapade

TASS, 2015. Otdel'naya Chast' Voisk Informatsionnykh Operatsij Poyavitsya Osen'yu v Krymu [A Separate Division of the Information Operations Troops will Appear in the Fall in Crimea]. *TASS*, 17 April. Available from: https://tass.ru/armiya-i-opk/1911074

The Moscow Times, 2023. Wagner Founder Prigozhin Admits He Was Behind Russia's Infamous Troll Farms. *The Moscow Times*, 14 February. Available from: www.themosc owtimes.com/2023/02/14/wagner-founder-prigozhin-admits-he-was-behind-russias-infamous-troll-farms-a80228

Thornton, R. and Miron, M., 2020. Towards the "Third Revolution in Military Affairs": The Russian Military's Use of AI-enabled Cyber-Warfare. *RUSI*, 165(3), 12–21. https://doi. org/10.1080/03071847.2020.1765514

Tucker, E., 2023. US Busts Russian Cyber Operation in Dozens of Countries. *The Associated Press*, 9 May. Available from: www.militarytimes.com/news/your-military/2023/05/09/ us-busts-russian-cyber-operation-in-dozens-of-countries/

UK Government, 2022. UK Exposes Sick Russian Troll Factory Plaguing Social Media with Kremlin Propaganda. 1 May. Available from: www.gov.uk/government/news/uk-expo ses-sick-russian-troll-factory-plaguing-social-media-with-kremlin-propaganda

Urban, E., 2022. Cyber-Attacken auf die Ukraine: Wiper-Malware Befällt "Hunderte Computer." *T3N*, 24 February. Available from: https://t3n.de/news/cyber-attacken-ukra ine-wiper-malware-1454318/

Yang, A., 2019. Reflexive Control and Cognitive Vulnerability in the 2016 U.S. Presidential Election. *Journal of Information Warfare*, 18(3), 99–122.

Yourish, K., Buchanan, L. and Watkins, D., 2018. A Timeline Showing the Full Scale of Russia's Unprecedented Interference in the 2016 Election, and Its Aftermath. *The New York Times*, 20 September. Available from: www.nytimes.com/interactive/2018/09/20/us/politics/russia-trump-election-timeline.html

Zetter, K., 2016. Inside the Cunning, Unprecedented Hack of Ukraine's Power Grid. *Wired*, 3 March. Available from: www.wired.com/2016/03/inside-cunning-unprecedented-hack-ukraines-power-grid/

Zhora, V., 2022. How to Ride a Bear – Russian Cyber Posture and Security Implications. *In: The CyberSec Forum/Expo*, 18 May 2022 Katowice, Poland. Available from: www.youtube.com/watch?v=ll7PQP_IcdA

5 Everyone a Sensor

The Implications of the Russo-Ukrainian War and the Democratization of Intelligence for Great Power Competition

David V. Gioe and Tony Manganello

Introduction: Intelligence Beyond the State

Throughout the history of interstate competition, governments have sought to utilize intelligence to gain an advantage over adversaries and to shape international narratives in their favor. The Cold War, in particular, provided a crucible for refining how to wield information within the context of superpower rivalry. In one oft-cited example from October 1962, Adlai Stevenson, the U.S. Ambassador to the United Nations, grilled Soviet Ambassador Valerian Zorin about whether the Soviet Union had deployed nuclear-capable missiles to Cuba (Lindsay 2012). While Zorin waffled (and did not know in any case), Stevenson went in for the kill: "I am prepared to wait for an answer until Hell freezes over ... I am also prepared to present the evidence in this room" (*U.S. ambassador Adlai Stevenson at the UN 1962 Cuban Missile Crisis* 2017). Stevenson then theatrically revealed several poster-sized photographs based on the U.S. U-2 spy plane collection showing Soviet missile bases in Cuba, directly contradicting Soviet claims to the contrary. It was the first time that (formerly classified) imagery intelligence (IMINT) had been marshaled as evidence to publicly refute another state in high-stakes diplomacy.

In the decades since Stevenson's theatrics, strategic revelation of classified information has been employed by the United States and Western allies to influence international narratives. The lead-up to the 2022 Russian invasion of Ukraine is the latest example in which Western partners openly discussed intelligence assessments of the mobilization of Russian forces, drawing global attention to the imminent invasion (Marrow and Vasovic 2022). Although the discussion of classified assessments in public did not deter Moscow from invading, the Russia–Ukraine war has provided useful lessons within the fields of open-source intelligence (OSINT) and great power competition (GPC). This chapter explores the impact of OSINT within the context of the Russia–Ukraine conflict, mining this case for implications for near-term GPC.

In Stevenson's day, the UN audience was stunned by the revelation of U.S. intelligence's high-altitude collection capabilities. During the Cuban missile crisis – and indeed through the end of the Cold War – such exquisite airborne and

DOI: 10.4324/9781003425304-6

satellite collection was exclusively the purview of the United States, the United Kingdom, and the USSR (Brands 2021). Due to vast (and rapid) technological advances, the world (and the world of intelligence) has democratized considerably in the past 60 years, adding more non-state (and private sector) actors to the intelligence field. By the time Russian President Vladimir Putin launched his so-called "special military operation" (Marshall 2022) in Ukraine in late February 2022, imagery intelligence (IMINT) and geospatial intelligence (GEOINT) were already highly democratized intelligence disciplines (INTs). Commercial satellite companies such as Maxar and Google Earth now provide high-resolution images free of charge. Previously the sole domain of government intelligence services and military commands, high-resolution imagery is now available in the public-analysis domain in near real-time and commercial imagery products are now routinely requested by states (Janjeva et al. 2022). Thanks to such ubiquitous imagery in the public domain, anyone could see – in remarkable clarity – that the Russian military was massing on Ukraine's border in early 2022 (RFE/RL 2022).

Furthermore, the overhead collection is not the only domain experiencing democratization and impacting the course of this conflict. Ford and Hoskins observe that due to the proliferation of smartphone technology, regular people are now part of the "machinery of war" in a "convergence between the weapons of war and the media of war" (2022). Due to the profusion of smartphones and other connected devices, everyone is now a potential intelligence collector, a development that is changing the character of warfighting. The Russia–Ukraine conflict has become a prime case study of this convergence.

Geolocation-stamped photos and user-generated videos uploaded to social media platforms such as Telegram (Bergengruen 2022) or TikTok (Sonne et al. 2022) enabled further refinement of – and confidence in – the democratized view of Russian military activity. And continued citizen-driven collection and analysis showed changes in Russian positions over time without needing to wait for another satellite to revisit the area. Of course, such a display of force was not guaranteed to presage an invasion, but there was no hiding the composition and scale of the buildup – and the eventual invasion came as a surprise to no one.

Once the Russians actually invaded, another key development ensued: the democratization of near real-time battlefield awareness. In a digitally connected environment, everyone can be a sensor or intelligence collector, wittingly, or unwittingly. Furthermore, hobbyists the world over can now develop insightful analyses based on the quality and quantity of publicly available information. This dispersed and crowd-sourced collection against the Russian campaign was assisted by the huge number of people taking pictures of Russian military equipment and formations in Ukraine and posting them online. These average citizens likely had no idea what exactly they were snapping a picture of but established military experts on the internet do. Sometimes, within minutes internet platforms like Twitter/X had threads and threads of what the pictures were and what they revealed, providing what military intelligence professionals call the Russian order of battle.

These and similar developments within the information space are drastically and swiftly altering the face of war and carry important implications for GPC. The

Russia–Ukraine conflict has become an instructive lens through which to analyze the near-future terrain of GPC, which will have spillover effects far beyond Eastern Europe. For instance, in February 2023, the U.S. employed intelligence assessments to publicly warn China of the political ramifications of providing weapons and ammunition to Russia (Barnes and Entous 2023), a signal that the dynamics of the Ukraine war may impact future U.S.–China behavior toward Taiwan (Singleton 2023). Using classified intelligence to undermine GPC competitors like Russia and China has been part of the U.S. intelligence playbook since the Cold War, but new areas of open-source collection through publicly available information may have an even larger impact. States that make investments in creating asymmetries in the information space stand to gain intelligence advantages in the current era of GPC. This chapter will now examine the case of Russia–Ukraine to demonstrate recent developments in democratized intelligence and their implications for the GPC landscape. Dividing a discussion of OSINT activities into collection and analysis, we examine the development of OSINT-driven impacts on these two core functions of intelligence. Finally, we conclude with a discussion of how these developments may impact the near-future terrain of GPC.

Definitions

The intelligence literature is awash in terms relating to the information domain, and a brief primer on how this chapter makes use of some of the common terms is in order. "Open source intelligence (OSINT)," "publicly available information (PAI)," "information warfare (IW)," and "information operations (IO)" are all widely employed by an array of authors, sometimes with confusing or conflicting meanings. We begin with our understanding and application of these often vague terms.

Firstly, intelligence professionals understand the terms "intelligence" and "information" as differentiated from one another. Generally, "information" is what is collected by intelligence collection platforms, epistemically categorized by the intelligence discipline (INT) that produced them – signals intelligence (SIGINT) intercepts, human intelligence (HUMINT) reports, overhead imagery (IMINT or GEOINT), and so on. "Intelligence" refers to a broader set of interpretations (and, in some cases, applications via covert action) derived from the collected information and integrative analysis performed by experts for a purpose. Thus, intelligence is estimative in nature, not a declaration of unassailable truth (Lowenthal 2023). By extension, OSINT, as an intelligence discipline, implies that a level of interpretation or assessment is required.

As a conceptualization, OSINT is not only the information collected via open (or sometimes "grey") sources via PAI, but also the interpretation of what the PAI data mean, and how their analysis impacts political, strategic, and tactical decisions or other targeted outcomes (Janjeva et al. 2022). PAI is also a broad term and can mean information available for free or behind paywalls or other password-protected areas (sometimes referred to as "grey" sources). Therefore, all PAI is not created equal, and should be understood within a spectrum of gradations. However, for simplicity, intelligence professionals think of PAI as any information not deriving from classified sources and OSINT as a collection and analytical discipline which integrates PAI – the results of which, if produced by a government intelligence service, are usually classified.

The operationalization of OSINT can be referred to as information warfare (IW) or information operations (IO) when integrated into a specific (but usually broader) military or political objective. These terms are often used interchangeably, but getting too granular is not needed here. However, it is useful to note that both share common purposes: to create asymmetric advantages and to influence public narratives, and sometimes private ones too. Along with discussing the collection and analytical dimensions of OSINT, this chapter also uses examples of IW and IO effects to illustrate the dynamics of the information domain.

Democratized Intelligence Collection

The democratization encouraged by the diffusion of technologies such as social media and commercial imagery has created analytically useful intelligence collection asymmetries in the Russia–Ukraine conflict. To give an example of how this has been playing out, according to one media account, "[i]n one video posted to TikTok on February 5, a man walking his dog in a Russian town a few hours' drive from the Ukrainian border captured missiles passing by on a snowy street" (Sonne et al. 2022). According to the report, a Russian military analyst at the RAND Corporation provided confirmation that the observed munitions were Iskander ballistic missiles, a type of missile expected to be deployed against strategic targets (Sonne et al. 2022). Since then, the Iskander has been used repeatedly in the war, for instance, targeting a railway station and killing dozens of people on Ukraine's independence day (Thebault and Hendrix 2022). This fusion of PAI and public-domain analysis has been an emerging and enduring feature of the conflict to date, with serious battlefield (and beyond) implications.

The "loose lips sink ships" bromide has always been employed by security professionals to preserve operations security (OPSEC) and defeat hostile collection attempts, however, embedded geolocation data in commonly used apps and devices has qualitatively changed what constitutes "lips." One of the early revelations of how publicly available geolocation data can be used to reveal sensitive information happened in 2018. The fitness tracking app Strava published a world "heat map" which overlayed human patterns of movement onto maps based on data generated by fitness trackers such as Fitbits. As hobbyists and professionals analyzed the publicly available Strava heat map, brightly outlined shapes appeared in remote and dangerous areas, revealing the perimeters of probable military locations in hot spots such as Afghanistan, Djibouti, and Syria (Hern 2018). The bright lines detailed the exercise routines and walking paths of U.S. military members stationed at sensitive locations, effectively mapping the military installations for all to see. Militaries and intelligence services the world over were put on notice about the unwitting OPSEC dangers of app-generated data.

Despite the demonstrated risk location data pose to OPSEC, Russian soldiers themselves have committed grave blunders in the Ukrainian theater. For example, in December 2022, a Russian soldier posted photos and videos of his unit, with geolocation data enabled, allowing Ukrainian forces to pinpoint his location at a country club housing Russian troops. On December 20, that country club was destroyed, and subsequent photos appeared on the same Russian soldier's social media feed, showing the destruction and in the process, providing Ukraine with an accurate battle damage assessment to boot (Schogol 2023). At a time when Russia is considered, by some, a deft operator in the information domain, lapses like this are surprising. No doubt Russian IO/IW planners are translating these experiences into future strategies.

Alternately, Ukrainian forces seemed to have anticipated both the opportunities and pitfalls of social media-generated data, effectively gaining both IW and battlefield advantages in some cases. For example, Ukrainian commanders have apparently accepted the counterintelligence and security risks of soldiers using mobile devices in a tactical environment in order to broadcast images of the Russian invasion to the wider world. However, these battlefield posts appear to have been curated as no footage that might negatively impact perceptions (such as the mistreatment of Russian captives) has surfaced (Kleisner and Garmey 2022). This level of anticipation, and the skillfulness of execution, indicate that the Ukrainians have a sophisticated IO/IW campaign underway (O'Brien 2022). By contrast, Russian soldiers have been posting apparently unfettered – to devastating tactical effect.

In this conflict, soldiers are not the only people producing tactically useful user-generated content. Such a web of seemingly infinite data and metadata (often referred to as "digital exhaust") can also be emitted both intentionally and unintentionally by non-combatants and then operationalized by the military, who is both mining for such digital treasures and who stands ready to respond. For instance, the social media posts of Russian holidaymakers in Crimea and careless journalists have been exploited by Ukrainian forces. In one post, a Russian man on vacation posed for a photograph in front of a Russian S-400 Air Defense system on a beach in Crimea, revealing its location for Ukrainian forces (who mocked him for the mistake) (Backhouse 2022). In another instance, a careless Russian war propagandist posted a picture of himself at the Wagner Group headquarters in Popasna, Ukraine, and accidentally revealed a street sign (Soteriou 2022), reportedly leading to a devastating artillery strike on Putin's mercenaries (Sreda 2022). Such cases illustrate the challenges and opportunities of social media in the PAI realm.

Far beyond mere advances in the availability and quality of imagery and geolocation data, ubiquitously available data (and curated datasets) from commercial geospatial intelligence operators include a growing array of technological collection capabilities. Such capabilities include highly technical intelligence collection platforms such as synthetic aperture radar (SAR) satellites, which provide a view at night and through clouds, radio frequency (RF) monitoring, and GPS jamming (Datta 2022). Meanwhile, NASA's Fire Information for Resource Management System (FIRMS) satellite data, designed to track wildfires, is being used to identify firefights in Ukraine from space. There is even democratized signals intelligence, in which citizens can tune into unencrypted Russian military communications and then create Twitter (now X) thread (ShadowBreak Intl. [@sbreakintl] 2022) with translation and analysis, revealing much about Russian losses and soldier morale (Horton and Harris 2022). The availability of copious PAI has spawned a cottage industry of public-domain intelligence collection and analysis.

Furthermore, open-source researchers and digital activists are also identifying and debunking false claims and exposing disinformation (Moran 2022; Verma 2022), and documenting human rights abuses (Koenig 2022). These contributions to a robust and multidimensional understanding of the Ukraine war both bring distant observers meaningfully nearer to the conflict and enhance the credibility of official investigations when the findings track closely with one another. There is, of course, a spectrum here, and not all PAI is free. Private companies such as *the Bellingcat* purchase "grey" data and use it in their analysis, which is freely available, and joins a growing list of other open-source investigative offerings such as the Institute for the Study of War and the *New York Times*. The upshot is that increasingly abundant, detailed, and technical data are available through PAI, providing the publics with unprecedented optics (and influence) into battlespace realities that have historically been tightly controlled views manipulated by the combatants. Turning now to how PAI data is used by state and non-state actors, we discuss the analytical outputs of the OSINT environment.

Beyond Collection: Democratized Intelligence Analysis

Collection, of course, is only a part of the intelligence process. Analysis – interpreting meaning and generating useful insights from collected information – has also become increasingly democratized and nongovernmental experts online have provided astoundingly useful insight about the successes and failures of various phases of the Russian campaign.[1] The average person almost anywhere in the world now has available to both bird's eye and human points of view of the battlespace as well as analytical assessments that previous military commanders would covet. Indeed, it is likely that academic and think tank assessments of how the war is unfolding are not that different than the briefings for Western generals and policymakers.[2]

Crowdsourced data aided by commercially available mapping tools (and algorithms) challenge the very notion that only secret intelligence can narrow the cone of uncertainty for intelligence consumers. For instance, amateurs on Twitter discovered "traffic jams" on the Russian side of the border using Google Maps, right before the 24 February 2022 invasion (The Economist 2023). Formerly the purview of secret government agencies, private sleuths can now interpret data and make analytical models to show, for instance, where artillery is being fired from with precise locations (Case 2016). In one case, Dutch open-source intelligence outfit Oryx has meticulously documented Russian order of battle losses in Ukraine based on imagery analysis and social media reporting (Mitzer and Janovsky 2022).

These examples illustrate the power of joining information generated from a nonexpert human sensor on the ground with a military analyst thousands of miles away, connected by online discussion forums that would not seem out of place in a military intelligence agency. Meanwhile, governments are pushing their own analytical outputs into the public square via the Internet. The UK's Ministry of Defence (MOD) and the Ukrainian MOD routinely tweet analytical updates, which can be integrated by public intelligence organizations. The rise of independent intelligence analysis appearing throughout the past year of this conflict presages a model of public-domain analysis likely to proliferate further in future conflicts.

Democratization of the information environment has allowed experts and hobbyists alike to join across the internet using (mostly free) data and tools to generate perspectives that, heretofore, would have required significant investments of time and resources by government agencies. Even beyond battlefield awareness in Ukraine, citizen-activists (in one case, a teenager) have been tracking the whereabouts of Russian oligarchs' conveyances, creating a mosaic that includes elite Russian activity far beyond the theater of the war (Valinsky 2022). Even the movements of billionaires like Elon Musk are now easily followed via open-source flight tracking sites that utilize the tail numbers of private aircraft to reveal their flight plans publicly (Chang and Masunaga 2022). The resulting public-domain analysis has hinted at potential moves within the trillion-dollar enterprises represented by players like Musk, causing some to employ counterintelligence tradecraft to protect proprietary information (Chang and Masunaga 2022). If titans of industry are compelled to employ counterintelligence strategies based on public analysis, Russian oligarchs will almost certainly take steps to conceal their activities as well. Perhaps OSINT investigations will reveal security gaps that astute intelligence services will seek to close by improved denial and deception – a potential unintended consequence of a more democratized information environment.

Furthermore, democratized intelligence is not just happening in parallel with professional intelligence agencies, they are interacting with each other, too. Private analysts are using the government information and either confirming or refuting the analysis, while government analysts are finding increasing utility in what is available from outside their secure confines – in some cases allowing professional intelligence services to cite publicly produced intelligence while maintaining the secrecy of their own sources and methods. The hastening democratization of the information environment – as clearly demonstrated by the Russia–Ukraine conflict – has allowed citizens to seize agency by sharing real-time battlefield information and analysis, greatly reducing the historic monopoly on combatants controlling the prevailing narratives of the theater of operations (Watling 2021).

However, not all public-domain analysis has the same merit – nor should be given equal weight in assessments by state intelligence bureaus. Some caution is needed in assessing the value of public-domain OSINT outputs. For starters, independent analysts do not operate along the contours of the formal intelligence cycle. Many are informal and self-tasked – they do not operate from strategic planning requirements like government intelligence services are required to do. Furthermore, lacking professional standards, controls, and training, amateur analysts (and collectors) could put people in real danger or be used to unwittingly spread disinformation (Verma 2022). Without accountability (and proper attribution), public-domain analysis could be weaponized by a sophisticated adversary. This downside to public-created/disseminated intelligence will need to be mitigated by government-run enterprises as we consider what OSINT developments portend for GPC.

Future Implications of Democratized Intelligence for GPC

Peering ahead, in future wars combatants should expect to be confronted with this changing nature of secrecy hindering operational security and making surprise much more difficult to achieve. State efforts to spin or control the battlefield narratives will also be challenged, likely forcing combatants to react to reliable and convincing evidence-based analysis from the public-analysis realm. To illustrate just how quickly this topic is developing, consider that in his 2014 invasion of Crimea, Putin was able to control much of the narrative and create confusion until the annexation was accomplished as *fait accompli*, but in 2022, he was unable to spin the false narratives that he was relying upon to justify his invasion. While Anglo-American intelligence agencies declassified troves of intelligence to "pre-bunk" Putin's ploys (Myre 2022), private organizations like *the Bellingcat* were also on the scent (Dreger 2022). This evolution took less than a decade, catching an occasionally crafty actor like Putin flatfooted. As great power competitors glean lessons learned from the Russia–Ukraine experience, future planners will be less likely to be caught off guard by this loss of message control.

Not unlike the way that American Ambassador Stevenson was able to confront the Soviet Ambassador Zorin during the Cuban missile crisis with irrefutable evidence, individuals harnessing PAI have been able to undermine Russian narratives with convincing interpretations of available crowd-sourced intelligence. Future combatants or competitors will contend with an increasing level of public visibility into their actions and must plan accordingly. Like Wikipedia, much of the work is self-policed. Even when the nongovernmental experts disagree, their critiques of each other's work sharpen the analytical picture for everyone – laying bare capabilities and intentions in a way previously unthinkable in past conflicts. Governments will need to accept and adapt to this new reality or operate at a disadvantage against those who have. The grip on the control of intelligence-informed conflict narratives by states is loosening and not likely to be regained.

The future implications of democratized intelligence collection and analysis will shape the conduct of war since the mountains of data and ways to collect it will only continue to accelerate and increasingly outpace government intelligence services' capabilities to process and analyze so much data. While technology has allowed machines to collect and process data at exponentially greater rates, the cognitive capacity of human analysts has not qualitatively changed. Western intelligence communities will need to reckon with this new phenomenon as they will be unable to put this genie back in the bottle. Intelligence communities will likely face pressure from their political leaders and publics to keep up with what the outside experts are saying in terms of increasingly transparent public engagement (and also feel pressured to show that they are, in fact, keeping up as a point of professional pride). Nevertheless, intelligence communities may again feel the longstanding tension between doing strategic intelligence and also trying to be speedy and agile to craft tactical assessments as quickly as external observers have demonstrated that they can do well. Past generations of intelligence professionals routinely felt the pressure to "beat the press" in reporting to policymakers. Now, intelligence shops will feel pressure to keep pace with a diffused network of independent public-domain analysts armed with increasingly available high-quality data and far fewer internal coordination processes to contend with.

Moreover, readily available and highly reliable information has enabled intelligence agencies to be less secretive by pointing, in public, to outside assessments that track with their own internal assessments (Dylan and Maguire 2022). In a sense, this enables the public a peek behind the classified curtain, and it is enabled by democratized intelligence. For example, in previous generations, government-imposed "shutter control" could limit the amount and quality of commercially available imagery (Lowenthal 2023). While there are still government-imposed limits on commercial imagery capabilities, this monopoly has been largely broken and will continue dissipating. As Philip H.J. Davies observed, there is a convergence happening between open-source intelligence and open government (British Academy Conference Presentation, November 2022). These synergies seem to auger well for public understanding of the vital role of secret intelligence in democratic societies. Furthermore, OSINT cutting both ways – classified data being made public, and public-domain creation of OSINT – will be a permanent feature of the intelligence landscape moving forward.

The first Gulf War (1990–1991) was often called the first TV war in which television cameras brought the war into people's living rooms on a nightly basis (Johnson 2015). The Russian invasion of Ukraine may be the first "social media war," dramatically shortening the space between sensor and public awareness, and the reverberations will echo into future conflicts. We can already discern how the information domain might define the contours of the U.S.–China competition taking shape over potential flashpoints, like Taiwan (Singleton 2023) and Chinese territorial claims in the South China Sea. This democratization of intelligence, which produces an unprecedented level of battlefield transparency to the public, will change how wars are conducted, how they are reported, how they are monitored and assessed, how war crimes are documented, and also, perhaps who counts as a participant in the conflict itself. In sum, these developments represent nothing less than a technologically enabled step change in intelligence, in warfare and may herald the new face of war itself. Intelligence scholar Amy Zegart has observed, "In today's world of information warfare, weapons don't look like weapons" (2022). In such a world, the legacy intelligence services of great power nations will need to adapt to the accelerating changes within the information environment.

Countries that fail to appreciate the role of OSINT and make investments in creating asymmetries will be at a disadvantage. Which investments to make (AI/ML and other big data processing efforts, for example) and deciding how to structure OSINT efforts are key areas of concern for decision-makers. In the United States, for example, dueling schools of thought will need to be settled at a philosophical level. Zegart, among others, has advocated for creating a dedicated, independent U.S. OSINT intelligence agency to keep pace with competitors and avoid calamitous surprises (2022). On the other hand, Joseph Hatfield (2023) compellingly argues that the term "OSINT" itself is no longer a useful concept, consigning it to "junk drawer" status and calling for a "dehomogenization" of OSINT, splitting it out into the taxonomies of the traditional source-based intelligence disciplines – imagery whether commercial or government-generated is IMINT, and so on. Moreover, the integration of OSINT efforts across Western services poses its own challenges. For example, a June 2022 UK assessment eschews the idea of establishing an independent OSINT agency within the British intelligence community (Janjeva et al. 2022).

Such debates about how to organize government use and understanding of OSINT will need to settled sooner than later as the information space is already a tensely contested domain of GPC. However, decisions like these are not new territory for intelligence communities. For instance, after the creation of the Office of the Director of National Intelligence (ODNI), structural questions arose as the historically geographical orientation of intelligence activities made room for functional centers organized around transnational issues such as terrorism and proliferation. Similar decisions will need to be made regarding OSINT. The old analogy of constructing an aircraft already in-flight comes to mind.

Conclusion: Democratized Intelligence Matters for GPC

For conflicts that result in war over physical territory, violence will remain a key aspect. Military planners will always need to look for ways to develop superiority within the physical (and virtual) domains of warfare. However, in a rapidly changing and democratizing information environment, influencing the shape of international narratives will become more of a factor in GPC, with states like China and Russia already demonstrating prowess to employ sophisticated influence operations (Ryan 2022). Furthermore, PAI and OSINT will continue to impact battlespace awareness and operations. In an age where technological diffusion has made everyone a sensor, no future battlefield narratives will be under the sole influence of the combatants. As demonstrated by the Russia–Ukraine war, public domain collection and analysis has proven to be a disruptor to the traditional monopolies states enjoyed over the intelligence domain of warfare and have led to asymmetries in the warfighting capabilities of the two sides – and, after nearly two years of fighting, has contributed to the results so far.

However, unlike war, which is perceived by some states as binary (nations are either at war or at peace), GPC exists in a continuous state. In the near term, a Western power is unlikely to go to war with another great power nation directly, but in an environment of persistent GPC, OSINT will play an ascending role. As Western democracies look to the future of GPC with aggressive revisionist states like Russia and China, understanding the world of PAI and harnessing the benefits of OSINT will become key components to creating decision advantage – the core mission of intelligence services. Whether creating independent OSINT agencies or better integrating OSINT efforts within existing structures, Western intelligence communities will need to make large investments of resources (and shift their culture) into the information environment. China, Russia, and other adversaries have made their intentions to dominate this domain clear (*Annual Threat Assessment of the U.S. Intelligence Community* 2023). In the deluge of technological shifts, such as the phenomenon of democratized intelligence, as explored in this chapter, opportunities will arise to create asymmetric advantages in the GPC arena. Those best prepared to harness the opportunities and avoid the pitfalls will have a clear edge in struggles yet to come.

Notes

1 Prominent examples include Harvard University (n.d.), King's College London (n.d.), and Lawrence Freedman (n.d.).
2 For a good example, see the Institute for the Study of War (2022).

References

Annual Threat Assessment of the U.S. Intelligence Community, 2023. Washington, DC: Office of the Director of National Intelligence.
Backhouse, A., 2022. Tourist in Crimea Accidentally Exposes Russian Military Position in Shirtless Vacation Photos. *New York Post*, 25 August.
Barnes, J. and Entous, A., 2023. How the US Adopted a New Intelligence Playbook to Expose Russia's War Plans. *New York Times*, 23 February.
Bergengruen, V., 2022. Telegram Becomes a Digital Battlefield in Russia-Ukraine War. *Time*. Available from: https://time.com/6158437/telegram-russia-ukraine-information-war/ [Accessed 13 January 2023].
Brands, H., 2021. The Eurasian Century, Part IV: Cold War. *Engelsberg Ideas*, 6 December. Available from: https://engelsbergideas.com/essays/the-eurasian-century-part-iv-cold-war/
Case, S., 2016. Putin's Undeclared War: Summer 2014 – Russian Artillery Strikes against Ukraine. *Bellingcat*. Available from: www.bellingcat.com/news/uk-and-europe/2016/12/21/russian-artillery-strikes-against-ukraine/ [Accessed 2 March 2023].
Chang, A. and Masunaga, S., 2022. For the Rich and Famous, Private Jets are No Longer Private Enough. *Los Angeles Times*, 21 December. Available from: www.latimes.com/business/story/2022-12-21/la-fi-private-jet-tracking [Accessed 10 March 2023].
Datta, A., 2022. GeoInt, OSINT Comes Off Age For Near Real Time Coverage of Ukraine Conflict. *Geospatial World*, 7 March. Available from: www.geospatialworld.net/blogs/geoint-osint-comes-off-age-of-ukraine-conflict/
Dreger, G., 2022. Documenting and Debunking Dubious Footage from Ukraine's Frontlines [online]. *Bellingcat*. Available from: www.bellingcat.com/news/2022/02/23/documenting-and-debunking-dubious-footage-from-ukraines-frontlines/ [Accessed 6 March 2023].
Dylan, H. and Maguire, T. J., 2022. Secret Intelligence and Public Diplomacy in the Ukraine War. *Survival*, 64 (4), 33–74. https://doi.org/10.1080/00396338.2022.2103257

Ford, M. and Hoskins, A., 2022. *Radical War: Data, Attention and Control in the Twenty-First Century*. Oxford: Oxford University Press.

Freedman, L., n.d. Comment is Freed. *Substack* [blog]. Available from: https://substack.com/profile/69709932-lawrence-freedman

Harvard Kennedy School, n.d. Russia-Ukraine War: Insights and Analysis. *Harvard University*. Available from: www.hks.harvard.edu/russia-ukraine-war-insights-analysis

Hatfield, J.M., 2023. There Is No Such Thing as Open Source Intelligence. *International Journal of Intelligence and CounterIntelligence*, 1–22. https://doi.org/10.1080/08850607.2023.2172367

Hern, A., 2018. Fitness Tracking App Strava Gives Away Location of Secret US Army Bases. *The Guardian*, 28 January. Available from: www.theguardian.com/world/2018/jan/28/fitness-tracking-app-gives-away-location-of-secret-us-army-bases [Accessed 6 March 2023].

Horton, A. and Harris, S., 2022. Russian Troops' Tendency to Talk on Unsecured Lines Is Proving Costly. *The Washington Post*, 28 March. Available from: www.washingtonpost.com/national-security/2022/03/27/russian-military-unsecured-communications/

Institute for the Study of War, 2022. Russian Offensive Campaign Assessment, October 5. *Institute for the Study of War*, 5 October. Available from: www.understandingwar.org/backgrounder/russian-offensive-campaign-assessment-october-5

Janjeva, A., Harris, A., and Byrne, J., 2022. The Future of Open Source Intelligence for UK National Security. *RUSI Occasional Paper*, 7 June. Available from: https://rusi.org/explore-our-research/publications/occasional-papers/future-open-source-intelligence-uk-national-security

Johnson, T., 2015. Desert Storm: The First War Televised Live Around the World (and Around the Clock). *Atlanta Magazine*, 18 March. Available from: www.atlantamagazine.com/90s/desert-storm-the-first-war-televised-live-around-the-world-and-around-the-clock/

King's College London, n.d. The War on Ukraine Explained: Hear from Our Experts. *King's College London*. Available from: www.kcl.ac.uk/the-ukraine-crisis-explained-hear-from-our-experts

Kleisner, C.T.W. and Garmey, T.T., 2022. Tactical TikTok for Great Power Competition. *Military Review* (Online Exclusive), 1–20. Available from: www.armyupress.army.mil/journals/military-review/online-exclusive/2022-ole/kleisner-and-garmey/

Koenig, A., 2022. Seven Essential Questions for Ethical War Crimes Documentation. *Human Rights Center*, 1 June. Available from: https://medium.com/humanrightscenter/seven-essential-questions-for-ethical-war-crimes-documentation-6e891f498da6

Lindsay, J., 2012. TWE Remembers: Adlai Stevenson Dresses Down the Soviet Ambassador to the UN (Cuban Missile Crisis, Day Ten). *Council on Foreign Relations*, 25 October. Available from: www.cfr.org/blog/twe-remembers-adlai-stevenson-dresses-down-soviet-ambassador-un-cuban-missile-crisis-day-ten#:~:text=On%20Thursday%2C%20October%2025%2C%20the,as%20Americans%20watched%20on%20television

Lowenthal, M., 2023. *Intelligence: From Secrets To Policy 9e*. 9th ed. Thousand Oaks, CA: CQ Press/SAGE.

Marrow, A. and Vasovic, A., 2022. West Warns Military Build-up Near Ukraine Growing, Not Shrinking. *Reuters*, 16 February. Available from: www.reuters.com/world/europe/russian-pullout-meets-uk-scepticism-ukraine-defence-website-still-hacked-2022-02-16/ [Accessed 10 March 2023].

Marshall, T., 2022. Putin's new Iron Curtain – Will Democracies Stand up to Authoritarian States? *Engelsberg Ideas*, 24 February. Available from: https://engelsbergideas.com/notebook/putins-new-iron-curtain-will-democracies-stand-up-to-authoritarian-states/

Mitzer, S. and Janovsky, J., 2022. Attack On Europe: Documenting Russian Equipment Losses during The 2022 Russian Invasion of Ukraine. *Oryx*, 24 February. Available from: www.oryxspioenkop.com/2022/02/attack-on-europe-documenting-equipment.html

Moran, M., 2022. Open-Source Intelligence: How Digital Sleuths are Making Their Mark on the Ukraine War. *The Conversation*, 18 March. Available from: http://theconversation.com/open-source-intelligence-how-digital-sleuths-are-making-their-mark-on-the-ukraine-war-179135 [Accessed 2 March 2023].

Myre, G., 2022. As Russia Threatens Ukraine, the U.S. "Pre-bunks" Russian Propaganda. *NPR*, 8 February. Available from: www.npr.org/2022/02/08/1079213726/as-russia-threatens-ukraine-the-u-s-pre-bunks-russian-propaganda [Accessed 6 March 2023].

O'Brien, A., 2022. Open Source Intelligence May Be Changing Old-School War. *Wired*.

RFE/RL, 2022. In Photos: New Images Capture Russia Massing Weaponry Around Ukraine. *Radio Free Europe/Radio Liberty*, 10 February. Available from: www.rferl.org/a/satellite-photos-russia-ukraine-troop-buildup/31662944.html [Accessed 13 January 2023].

Ryan, M., 2022. *War Transformed: The Future of Twenty-First-Century Great Power Competition and Conflict*. Annapolis: Naval Institute Press.

Schogol, J., 2023. Russian Soldier Gave Away His Position with Geotagged Social Media Posts. *Task & Purpose*, 3 January. Available from: https://taskandpurpose.com/news/russian-military-opsec-failure-ukraine/

ShadowBreak Intl. [@sbreakintl], 2022. Intelligence Acquired Since the Beginning of the Russian Military Operation Over Ukraine Has Shown an Immense Lack of Logistic Support, Making This War One of the Most Unique in 2022 When It Comes to Surveillance. [A thread]. *Twitter*. Available from https://t.co/huILD8lEDX

Singleton, C., 2023. China's Ukraine Peace Plan Is Actually About Taiwan. *Foreign Policy*, 6 March. Available from: https://foreignpolicy.com/2023/03/06/china-russia-war-taiwan-ukraine-peace-plan-xi-putin/

Sonne, P., Lee, J. S., Ilyushina, M., Mellen, R., and Mirza, A., 2022. The TikTok Buildup: Videos Reveal Russian Forces Closing in on Ukraine. *Washington Post*, 11 February. Available from: www.washingtonpost.com/world/2022/02/11/russia-ukraine-military-videos-tiktok/ [Accessed 13 January 2023].

Soteriou, E., 2022. Ukraine Blows Up Russia's "Wagner Mercenary HQ" After Street Sign Gives Away Location. *LBC*, 15 August. Available from: www.lbc.co.uk/news/ukraine-wagner-mercenary-hq-location/ [Accessed 2 March 2023].

Sreda, S., 2022. "A Well-Aimed Strike" Ukrainian Forces Destroy Wagner Group Headquarters in Luhansk Region After Pro-Kremlin War Reporter Reveals its Location. *Meduza*, 15 August. Available from: https://meduza.io/en/feature/2022/08/15/a-well-aimed-strike [Accessed 2 March 2023].

Thebault, R. and Hendrix, S., 2022. Zelensky Vows Retribution After Deadly Russian Strike on Independence Day. *Washington Post*, 25 August. Available from: www.washingtonpost.com/world/2022/08/24/russia-ukraine-train-station-strike/

The Economist, 2023. Open-Source Intelligence Is Piercing the Fog of War in Ukraine. *The Economist*, 13 January. Available from: www.economist.com/interactive/international/2023/01/13/open-source-intelligence-is-piercing-the-fog-of-war-in-ukraine

U.S. Ambassador Adlai Stevenson at the UN 1962 Cuban Missile Crisis, 2017.

Valinsky, J., 2022. The Teenager Who Tracked Elon Musk's Jet is Now Tracking Russian Oligarchs. *CNN*, 2 March. Available from: www.cnn.com/2022/03/02/business/russian-jets-tracker-twitter-account/index.html [Accessed 10 March 2023].

Verma, P., 2022. The Rise of the Twitter Spies. *The Washington Post*, 22 March. Available from: www.washingtonpost.com/technology/2022/03/23/twitter-open-source-intelligence-ukraine/

Watling, J., 2021. Preparing Military Intelligence for Great Power Competition: Retooling the 2-Shop. *The RUSI Journal*, 166 (1), 68–80. https://doi.org/10.1080/03071847.2021.1923408

Zegart, A., 2022. Open Secrets: Ukraine and the Next Intelligence Revolution. *Foreign Affairs* (January/February 2023), 20 December. Available from: www.foreignaffairs.com/world/open-secrets-ukraine-intelligence-revolution-amy-zegart

6 In Africa, Great Power Competition Requires a Great Strategy for Information Operations

Tara Heidger and David Higgins

Introduction

The People's Republic of China (PRC) and the Russian Federation tend to be regarded together in sub-Saharan Africa as a monolithic Great Power threat against Western influence on the continent. However, each of these powers differs in their approaches and the sectors they target. Both countries seek to create a multipolar system in which they are the key players holding power and influence on the continent. To do this, Beijing relies on its economic weight while Moscow relies more on different forms of military cooperation and support. Both offer assistance free from the judgment and evaluation criteria that are required from Western engagement, and each also uses its influence on the UN Security Council to support its allies. China focuses on soft power, large-scale infrastructure projects, and associated financing. This builds dependency on Beijing, particularly in the face of debt servicing challenges resulting from years of over-borrowing and pandemic-induced economic losses. Russia, by contrast, has opted for a low investment/high yield strategy in which it uses arms sales, nuclear technology, and mercenaries to increase its influence, particularly in states already facing violent internal conflict.

Critically, however, in the past few years, the competition for influence in sub-Saharan Africa has evolved away from simple economic and military prowess, with Russia and China adept at influencing and gaining power in other heavy-handed regimes. President Xi Jinping is forthright about being opposed to democracy, human rights, and the free press. Meanwhile, President Vladimir Putin's February 2022 (re)invasion of Ukraine demonstrated to the world his lack of respect for democracy and provided valuable insights into his tight narrative control in Russia. This state-run media, in conjunction with other forms of economic and military assistance, has helped to build popular support across Africa for the invasion, with ties so strong that 24 African nations refused to denounce Russian aggression towards Ukraine at the United Nations in March 2022. What was an economic and military competition between states is now also one between authoritarian systems on the one hand and the democratic and economic prosperity of the West on the other, with African partnerships being the stage upon which this is played out.

DOI: 10.4324/9781003425304-7

Although our work as practitioners in the defense sector may mean we lack some theoretical appreciation of the complexities of the information landscape, what our work has provided is a broad understanding of the practical application of the information and influence activities being conducted by China and Russia on the African subcontinent. We suggest our practitioner's point of view will be helpful to those who find themselves needing to be able to identify and better understand the decisions being made by Russia and China while also appreciating the overarching objectives and effects of these decisions as they play out on the global stage.

People's Republic of China

The PRC prioritizes information activities in sub-Saharan Africa with the intent of increasing its standing globally. They do this by carefully leveraging their economic, political, and military relationships as required to guide the narrative. This section will discuss the ways China leverages the information environment by way of pro-PRC propaganda, purposeful dissemination of misleading content aimed at undermining Western narratives, and the application of censorship, suppression of unfavorable information about the PRC, and obstructing outlets that are critical of the regime (Cook 2021).

The best example of how China uses its activities in sub-Saharan Africa to influence narratives is the Belt and Road Initiative which represents over 1 billion people and covers about 20% of the Earth's landmass. Accompanying the one trillion-dollar infrastructure effort is a coordinated messaging and influence initiative led from Beijing, which simultaneously establishes a *new world media order*[1] (Radio Without Borders 2018) – allowing the PRC to effectively deflect any negative attention its way. In 2015 China launched the Digital Silk Road, and by early 2020 it was assessed that China had provided more financing for information and communications technology in Africa than all multilateral agencies and leading democracies combined (Arcesati 2020). PRC's influence in the information environment is not simply riding on massive numbers of Twitter accounts, bots, and carefully placed news coverage, but is built on a sustainable, holistic approach to modernizing its presence and control across the entire African continent. The Digital Silk Road includes far-reaching influence techniques including organizing international events with all-expense-paid trips for journalists to China, which allows for the promotion of its repressive vision of how the media should function while also allowing the PRC to tightly control the narrative. Reporters Without Borders published a report (Reporters Without Borders 2018) outlining how Chinese media outlets worldwide are beholden to the country's authoritarian leadership. In fact, this same organization ranked China 177th out of 180 in its 2021 Press Freedom Index. To undermine the West's best efforts to promote freedom of the press, Beijing actively encourages intimidation and violence to silence dissidents, even in democratic nations. From freelance reporters to major media outlets, from publishing houses to social media platforms, no link in the news production chain is immune to the "invisible hand" of Beijing. Democracies are struggling to react in the face of these threats. The media's lack of objectivity can often be found carrying over to international messaging to bolster the overarching Chinese information strategy. This expansion – the scale of which is still hard to gauge – poses a direct threat to emerging democracies across the African subcontinent.

China's narrative around infrastructure and economic development is typical of its foreign policy approach in developing countries (Ferchan 2020). Building a global information network serves Beijing's economic, security, and diplomatic goals by inaccurately reporting on topics that could harm China's political and economic relationship with the countries concerned. Foreign investment allows China to use its media and messaging apparatus to amplify infrastructure development efforts in Africa while covering up the many negative aspects of its own internal affairs. China's approach has supported the narrative in Africa that the Chinese are good partners for investment and infrastructure while also ignoring the many real problems that the Chinese Communist Party could present if African nations are brought closer to its authoritarian style of leadership.

China's information, communications, and technology investment abroad also allows for increased opportunities for companies such as Huawei to corner African markets. Huawei 5G phone networks and equipment have a reputation for being cheap and robust, and because the company has long been investing in remote areas, it has become an indispensable partner for many governments (Prinsloo 2020). Despite Western moves to interfere in the use of the Chinese mobile giant, many African countries continue to embrace it. Additionally, undisclosed terms within the memorandums of understanding between China and many African nations create increased opportunities for filtering, content moderation, data localization, and surveillance by China within the 5G and television networks. In Zambia, for example, Huawei reportedly provided technology to the country's increasingly authoritarian rulers to spy on political opponents – a technique that has been reportedly used within the government of Yoweri Museveni during recent Ugandan elections (Bariyo et al. 2019). China has cornered the market in Africa as the largest investor in physical and communications infrastructure, which means it is now in a stronger position to dictate and influence Africans and African governments' perceptions of and partnerships with China as a whole.

Influence through the Promotion of Soft Power

While the focus on Chinese efforts in Africa is often infrastructure-centric, there are also significant soft power inroads – focusing on social and human capital – being made by the PRC. The intent of this soft power is to cultivate a deeper understanding of China while reversing the Western hostility toward the country's development model and helping to bolster Beijing's image deficit abroad. Beijing's efforts include creating and maintaining educational and institutional relationships with both citizens and leaders across the continent. From Chinese university scholarships for an estimated 80,000 Africans to joint military exercises which open doors to arms deals and strategic military cooperation – China is outpacing the United States and Europe in building relationships. More recently, this soft power expansion has happened at the community level as well. Across 46 African nations, Confucius Institutes are spreading exposure to Chinese culture, history, and language while simultaneously acting as platforms for surveillance, censorship, and state-controlled propaganda. Despite the dark underside, this straightforward African access to China is contrasted with the often-exclusive access to opportunities within the United States. The Young African Leaders Initiative and Fulbright scholarships are known to exclude ordinary Africans as they require English, high levels of educational attainment, and personal essays outlining significant accomplishments to be considered.

In addition to partnerships with the general population, China is also building relationships through United Front Work Department. This agency is controlled by the Chinese Communist Party and is responsible for coordinating influence operations outside of China specifically working to target and build relationships with foreign actors and states. These efforts are hard to track but are known to play an increasing role in China's broader foreign policy. It is precisely the nature of United Front work to seek influence through connections that are difficult to publicly prove and to gain influence that is interwoven with sensitive issues such as ethnic, political, and national identity, making those who seek to identify the negative effects of such influence vulnerable to accusations of prejudice (Bowe 2018).

Publicly, China is also building relationships through its Forum on China-Africa Cooperation (FOCAC), which takes place every three years, building networks and growing interpersonal connections between powerful PRC and African leaders. By comparison, the United States has not convened a comparable event since Obama's U.S.–Africa Leaders' Summit in Washington, DC, in 2014. FOCAC agenda items include promises on educational partnerships, training on disaster relief, information technology, and law enforcement. These face-to-face engagements normalize the Chinese partnerships across the continent, and the established relationships allow for close communications and easy reach back with the PRC as required. These efforts build on the existing global narratives of partnership with African nations and allow the PRC to leverage them as required to maintain their influence.

Despite these soft-power initiatives, public opinion on the role of China has not significantly improved. President Xi was recently quoted by the state-run news service, Xinhua, as saying, "We must focus on setting the tone right, be open and confident but also modest and humble, and strive to create a credible, lovable and respectable image of China" (Xi quoted in Myers and Bradsher 2021) Across the developed world, opinions of China are low, and Pew Research Centre research from 2019 still shows in Africa, the United States has an advantage in terms of popularity and preference on the world stage. Soft power efforts by the PRC – estimated at $10bn per year spent globally by the PRC (Albert 2018) – do not erase the perceptions and reality of China's heavy hand both within China and abroad. Stories of human rights abuses toward Uyghurs still do make the news in Africa, and high rates of indebtedness by China globally also make headlines.

Influence through Military Expansion

The development of infrastructure across Africa is not the PRC being altruistic but is only the first step of Beijing's long-term strategic intent to influence and grow its global superpower status. Beyond the economic gains that stem from aggressive loan contracts and the attempt to soften China's image in the region is another goal: to expand China's military presence across Africa.

At the end of 2021, Equatorial Guinea was saddled with debt from the Chinese-built Bata Port, which was financed with $2bn of Chinese loans in 2014. Declining revenues as a result of corruption, mismanagement, and slumping oil prices left the estimated debt to China at 49.7% of Equatorial Guinea's GDP. Like other contracted debt-trap projects, such as the Port of Mombasa in Kenya, the penalty for default is the takeover of the asset. Thus, Port Bata will be the second Chinese military base in Africa and the first along the Atlantic Coast, expanding PRC military access to countries on the Atlantic.

The Equatorial Guinea naval base being handed over to China news broke immediately after the FOCAC 2021 summit in Dakar, Senegal, pointing to deeper security conversations happening with the PRC at this event. Moving away from its traditional emphasis on infrastructure development, Beijing used the meeting to emphasize a new theme of building "a China-Africa Community with a Shared Future in the New Era." Within the framework of fostering this "China-Africa Community," FOCAC's Action Plan 2022–2024 calls for the strengthening of "the implementation of the China-Africa peace and security plan" aimed at supporting "the building of the African Peace and Security Architecture" (Tanchum 2021). This allows the PRC to achieve its vision of sharing its achievements with the world and telling its story by growing the idea of China as a great power.

The good intent that is built through infrastructure investments, coupled with soft power relationships and the normalizing of the Chinese presence and cultural initiatives, have set the conditions for developing democracies to eagerly reach out to China and ask for support. In 2018, President Xi announced the creation of a $60bn pot of Chinese money earmarked for development projects in Africa – including commercial seaports – and every single African nation understandably wants a piece of that pot. It is exactly that sort of initiative that buys the PRC the level of influence it requires to gain support for military cooperation, key infrastructure initiatives, and standing up against the United States and E.U. diplomacy on the continent. Other countries highly susceptible to Chinese military intervention include Angola, Seychelles, Kenya, and Tanzania, where China looks favorably to expand its African influence and control, cementing its ability to drive the narrative that they are a global force to be reckoned with.

Russia in Africa

Russian government policy (Office of the President of the Russian Federation 2021; Ministry of Foreign Affairs of the Russian Federation 2016) describes an international order characterized by a decline of the "West" alongside the rise of multiple alternative centers of geo-political power. Russia aims to play a leading role in such a "polycentric" world as a peer to the United States and China. However, despite the increase of Russian influence in Sub-Saharan Africa, Russia's lack of economic resources means that it cannot afford to achieve its ambitions on the continent in the same way as it might in its immediate periphery. Russia, therefore, uses information activities as a low-cost way to pursue its interests in Africa. This section describes the background to the Russian presence and why Russia is interested in the continent. It will then highlight the gap between President Putin's rhetoric and the level of resources relative to Russia's national ambition. Finally, it will cover the key information activities that Russia uses on the continent and how these support Russia's strategic objectives in Africa.

As part of its ideological struggle, the Soviet Union had extensive influence networks in Africa. These were largely based on its support for communist liberation movements in countries such as Angola, Mozambique, and South Africa (Department of State 1976, pp. 1–4) and socialist client states such as Ethiopia (Goodman 1990, pp. 5–8). The Soviet Union's collapse forced an economically straitened and politically chaotic Russia to abandon its international relationships and presence. After a period of minimal engagement, renewed stability allowed Russia to reassert itself on the African continent. Like the Soviet Union, Russia tried to contrast the absence of a colonial legacy in Africa against accusations of Western imperialism. However, lacking the Soviet Union's ability to accompany the spread of socialist ideology with huge amounts of aid, Russia has had to adopt a more transactional approach to its relationships.

Kremlin policy statements have thus adopted a more practical tone compared to liberation-era and postliberation Soviet sentiments. On the eve of the 2019 Russia-Africa Summit in Sochi, President Putin described Africa as "one of Russia's foreign policy priorities" (Putin 2019). In a pre-summit interview, he described Russian assistance to Africa as coming free of conditions, with Russia offering an alternative to the Western-led international order and proposing African solutions to African problems (Putin 2019). However, it is notable that President Putin's commitments were not then matched by economic reality. For example, whilst increasing rapidly in recent years to $20bn annually, Russian trade with sub-Saharan Africa only accounts for approximately 1% of global trade totals with the continent compared to $300bn with the European Union and $60bn with the United States (International Monetary Fund 2020).

Meanwhile, only 1% of foreign direct investment across the African continent originates from Russia (United Nations Commission on Trade and Development 2021). In addition to the low levels of economic resources that Russia has brought to bear, Russian foreign policy (Office of the President of the Russian Federation 2021) has strategically focused on Eurasia and Central Europe, with a longer-term shift towards China, India and the wider Middle East and Indo-Pacific region, paying little attention to Africa.

It is within this context that Russia has employed information activities to achieve a disproportionate influence relative to its resources. To achieve this, Russia aims to shape an African information order that allows it to promote its world view and to influence African elites' understanding of events in a way that supports its interests, builds its reputation, and wins the support of African leaders. It does this by attempting to influence the structure of the African information environment itself while also seeding narratives within it. This works on two levels; strategically, it aims to improve Russia's appearance as a global power while countering what it sees as a hostile global information environment that is structurally biased against its interests. At a more local level, Russia and its proxies such as Private Military Companies (PMCs) pursue more limited objectives such as improving consent for PMC presence and for other commercial activity.

To exert structural influence, Russia employs state-owned media outlets such as RT and Sputnik and other state actors such as the Ministry of Digital Development, Communications, and Mass Media to buy or otherwise support local media outlets. These co-opt local outlets by entering into a partnership and bilateral cooperation agreements with local state news agencies. For example, between 2015 and 2019, the Russian government and Russian state-affiliated media signed cooperation agreements with South Africa, the Republic of Congo, the Democratic Republic of Congo, Eritrea, and Cote d'Ivoire (Government Communications Information Service 2017; Ministry of Digital Development, Communications, and Mass Media of the Russian Federation 2019; RIA Novosti 2019; Ministry of Information 2019). The more visible outcome of these agreements is the dissemination of Russian content through state-backed news agencies (which, while official, may or may not themselves be trusted by the population of their countries). They also use measures such as relationship-building, joint projects, journalism training, and technical assistance (such as helping the Republic of Congo to switch to digital broadcasting) to embed Russian structures and Russian-leaning views into local information environments (Ministry of Digital Development, Communications and Mass Media of the Russian Federation 2019). This creates inter-governmental, institutional, and individual ties, outlooks, and ways of working, as well as pools of Russian-leaning journalists, enhancing the resilience of Russian influence and increasing its ability to withstand rival encroachment.

To seed narratives, RT and its more sensationalist sister state-owned news agency, Sputnik, form the core of the Russian state's media apparatus and make Russian-sponsored content available in English, Arabic, French and Spanish in almost every Sub-Saharan African country.[2] Individual narratives support specific views and policy objectives of the Kremlin but fall into three broad categories (NATO Strategic Communications Centre of Excellence 2021, pp. 6–7). Firstly, and most prominently, Russia takes advantage of its lack of a colonial presence in Africa to contrast itself with and subvert the positions of European powers. Secondly, it exaggerates the size and success of its activities and influence on the continent. Lastly, it emphasizes its assistance to African countries. Both outlets are seemingly successful in spreading large amounts of Russian content to support these narratives alongside positive news stories from Africa. This material is quoted, shared and discussed within local information environments including local or regional news sites, influential social media accounts and local political and civil society movements (Limonier 2019, p. 3).

This resharing allows RT and Sputnik to reach large and dispersed African audiences from French and English language global websites despite their lack of dedicated African channels and grounding in local information environments. It also helps to give them the appearance of legitimate news wires and to obscure the distinction with independent Western outlets such the BBC, CNN and Reuters (Simoyan 2012). However, even by the admission of RT's own editor-in-chief, Margarita Simoyan, RT instrumentalizes information as an extension of the Russian state rather than to inform viewership. Referring to the 2008 invasion of Georgia by Russia, Simoyan described RT in military terms: "The Ministry of Defence was at war with Georgia, and we waged an information war with the entire Western world," as well as explicitly stating that "there is no objectivity... So, when Russia is at war, we are, of course, on Russia's side" (Simoyan 2012). In this sense, RT and Sputnik are little more than a propaganda arm of the Russian state to serve specific foreign policy goals.

At a regional and country level, information activities are simultaneously employed to achieve Russian objectives directly and are also sold as a service alongside political advisers and mercenaries by state proxies such as PMCs. Many of the proxy organizations conducting these activities are owned or controlled (Harding and Burke 2019) by Putin associates and, until his death in August 2023, the so-called (Mackinnon 2021) Wagner Group Russian PMC network and Internet Research Agency "troll farm" financier, Yevgeny Prigozhin (Department of the Treasury 2016b). Perhaps the best example of this is in the Central African Republic (CAR), which illustrates how a relatively small deployment can have a disproportionate impact when targeting a fragile society with poor governance. Leaked Wagner network documents from 2018 indicated that cooperation between Russian PMCs and the CAR government was one of the highest among African countries. In fact, CAR received the joint highest rating amongst African countries in an internal Wagner assessment, scoring five out of five for the depth and scale of involvement, alongside Sudan and Madagascar (Harding and Burke 2019). The same documents said how CAR was "strategically important"; a "buffer zone between the Muslim north and Christian south"; that it would allow Russia to expand "across the continent"; and would enable Russian companies to strike lucrative mineral deals (Harding and Burke 2019). Russia uses such organizations to blur the line between the Russian government and the private sector, supported by the ambiguous use of military power.

Russia first deployed PMCs to CAR under UN Security Council Resolution 2127 (2013) to support the CAR Armed Forces (FACA) in their efforts to regain control of territory occupied by rebels, effectively lifting the embargo on the supply of arms to CAR (United Nations Central African Republic Sanctions Committee 2017). The Russian PMC operation has used sophisticated information activities to gain consent, or at least acquiescence, for their presence. The low levels of internet penetration has meant that many information activities are conducted through trad-itional media – (e.g., radio broadcast) and physical outreach. For example, the PMC Lobaye Invest has funded CAR media outlets, which reportedly suffer from a lack of advertising revenue, distributed leaflets that glorified the actions of President Touadera's MCU political party and Russia, while spreading anti-French and anti-colonial ideas (Dukhan 2020, p. 7), funded the radio station "Lengo Songo," whose website carries pro-Russian news but has no details about its backers, and in 2018 paid for the "Miss Centrafrique" beauty contest. Most recently, in March 2021, the UN Office for the High Commissioner of Human Rights published a damning report criticizing Russian PMC attacks on civilians (United Nations Office for the High Commissioner of Human Rights 2021). The report was followed in May 2021 by the premier in Bangui of the Russian film, "The Tourist" (Tourist Film Group 2021), which was dubbed into the local language Songo and 70,000 people attended. "The Tourist" was filmed in CAR in March and April 2021, and the opening credits said that it was inspired by the "heroic" Russian military instructors operating in CAR. This film appeared to have been backed by Yevgeny Prigozhin (Sauer 2021) and was heavily promoted by his RIA Fan media outlet (RIA Fan 2021). Prigozhin had close links to President Putin, owned Lobaye Invest, and was linked to sister PMC Sewa Security Services (Department of the Treasury 2020a), demonstrating the ambiguity of the relationship with the Russian state.

Prigozhin-linked entities have also conducted online information activities in Sub-Saharan Africa in support of Russian interests. In 2019, the Stanford Internet Observatory, working with the Facebook Threat Intelligence team, uncovered a huge online network targeting not just CAR but also the Democratic Republic of Congo, Madagascar, Mozambique, Sudan, and Libya. This network included fake news websites and pages purporting to belong to political parties and a politician in these countries, many of which were managed from Madagascar (Grossman et al. 2019, pp. 21–23). In CAR alone, Facebook detected 13 pages linked to Prighozin's companies. Nearly every page contained themes supporting President Touadéra, Russia and the Russian-brokered (but failed) peace agreement between the government and the rebels and criticism of the UN and France. Many pages shared content and similar posts appeared on multiple pages. Taken together, these initiatives have meant that despite Russian PMCs and associated individuals being sanctioned by the United States (Department of the Treasury 2020b) and criticized by the UN (United Nations Office of the High Commissioner for Human Rights 2021, p. 1) for human rights abuses in CAR, there is little interest in local media in these issues. Even very minor events such as the donation of a trampoline to a childcare center by Lobaye Invest have received more positive press coverage than the former (Patrick 2018). Russia instead receives positive reports (Huon and Ostrovsky 2018) in traditional and social media for its social and cultural activities, using websites including Russian-backed Facebook pages pretending to be native to CAR but managed either by Russians in CAR, or from a page owner in Madagascar (Grossman, Bush and DiResta 2019, p. 24; Proekt 2019). These fake sites actively promote narratives to unify target audiences behind President Touadera and his Russian partners; promote Russian support; disparage the UN and France; and to take defensive lines in response to criticism.

Conclusion

The United States and Western democracies stand to lose much of what has been gained over the past half-century should they fail to recognize that a significant portion of any "battle" is won in the information environment. To maintain positive Western perceptions in Africa, there needs to be increased and sustained Western partnerships and investment, more face-to-face interaction with African partners, and an exposing and undermining of the toxic relationship many countries are building and maintaining with the PRC and Russia. The Western intent in Africa must include convincing Africans and their leaders that in the ideological struggle between democracy and authoritarianism, democracy will prevail. The West must continually protect its interests and values on the continent by emphasizing that democracy is better, free markets are fairer, and human rights are protected through good governance and building a strong resistance to instability. The West must simultaneously focus its efforts on undermining the current Great Power Competition narratives. Exposing unfair practices, mis- and disinformation, and disrupting the reliance on the PRC supply chain will help contain the expansion and influence of authoritarian powers while deterring further actions and inroads for the PRC and Russia. The Western influence and information mission in Africa should repeatedly clarify for African leaders that partnering with, or choosing, authoritarianism over democracy is not the best long-term strategy.

Notes

1 China has been going to great lengths for the last decade to establish a "new world media order" under its control, with the aim of deterring and preventing any criticism of itself.
2 At time of writing: Angola, Benin, Burkina Faso, Burundi, Cameroon, Cape Verde, Central African Republic, Chad, Comoros, Cote d'Ivoire, Democratic Republic of Congo, Equatorial Guinea, Eritrea, Eswatini, Ethiopia, Gabon, Gambia, Ghana, Guinea, Kenya, Lesotho, Madagascar, Malawi, Mauritania, Mozambique, Namibia, Niger, Nigeria, Republic of Congo, Rwanda, Senegal, Seychelles, Somalia, South Africa, South Sudan, Sudan, Tanzania, Togo, Uganda, Zambia, Zimbabwe (RT n.d.).

References

Albert, Eleanor, 2018. China's Big Bet on Soft Power. *Council on Foreign Relations*, 9 February. Available from: www.cfr.org/backgrounder/chinas-big-bet-soft-power [Accessed 22 March 2022].

Arcesati, R., 2020. The Digital Silk Road is a Development Issue. *Mercator Institute for Chinese Studies*, April. Available from: https://merics.org/en/short-analysis/digital-silk-road-development-issue [Accessed 22 March 2022].

Bariyo, N., Chin, J., and Parkinson, J., 2019. Huawei Technicians Helped African Governments Spy on Political Opponents. *The Wall Street Journal*, 15 August. Available from: www.wsj.com/articles/huawei-technicians-helped-african-governments-spy-on-political-opponents-11565793017 [Accessed 12 March 2022].

Bowe, A., 2018. *China's Overseas United Front Work; Implications for the United States.* U.S.-China Economic and Security Review Commission. Available from: www.uscc. gov/sites/default/files/Research/China%27s%20Overseas%20United%20Front%20W ork%20-%20Background%20and%20Implications%20for%20US_final_0.pdf [Accessed 30 March 2022].

Cook, S., 2021. China's Global Media Footprint. *Sharp Power and Democratic Resilience Series. National Endowment for Democracy*, February. Available from: www.ned.org/ wp-content/uploads/2021/02/Chinas-Global-Media-Footprint-Democratic-Responses-to-Expanding-Authoritarian-Influence-Cook-Feb-2021.pdf [Accessed 20 March 2022].

Department of State, 1976. *National Intelligence Estimate.* Washington: Department of State. Available from: https://2001-2009.state.gov/documents/organization/70562.pdf [Accessed 3 January 2022].

Department of the Treasury, 2020a. *Cyber-Related Designations; Foreign Interference in U.S. Election Designations; Ukraine-/Russia-Related Designations.* Washington, DC: Department of the Treasury. https://home.treasury.gov/policy-issues/financial-sancti ons/recent-actions/20200923_33 [Accessed 19 February 2020].

Department of the Treasury, 2020b. *Treasury Targets Financier's Illicit Sanctions Evasion Activity.* Washington, DC: Department of the Treasury. Available from: https://home.treas ury.gov/news/press-releases/sm1058 [Accessed 23 January 2022].

Dukhan, N., 2020. *Central African Republic: Ground Zero for Russian Influence in Central Africa.* Washington, DC: Atlantic Council. Available from: www.atlanticcoun cil.org/wp-content/uploads/2020/10/CAR-Russian-Influence-Final.pdf [Accessed 24 January 2022].

Ferchan, M., 2020. How China is Reshaping International Development. *Carnegie Endowment for International Peace*, 8 January. Available from: https://carnegieendowm ent.org/2020/01/08/how-china-is-reshaping-international-development-pub-80703 [Accessed 30 March 2022].

Goodman, M., 1990. *Gorbachev and Soviet Policy in the Third World.* Washington, DC: The Institute for National Strategic Studies. Available from: www.files.ethz.ch/isn/23416/ mcnair06.pdf [Accessed 3 January 2022].

Government Communications Information Service, 2017. *South Africa and Russia to Share Communication Efforts.* Pretoria: Government Communications Information Service. Available from: www.vukuzenzele.gov.za/south-africa-and-russia-share-communication-efforts [Accessed 16 January 2022].

Grossman, S., Bush, D., and DiResta, R., 2019. *Evidence of Russia-Linked Influence Operations in Africa.* Stanford Internet Observatory. Available from: https://fsi-live. s3.us-west-1.amazonaws.com/s3fs-public/29oct2019_sio_-_russia_linked_influence_o perations_in_africa.final_.pdf [Accessed 24 January 2022].

Harding L. and Burke, L., 2019. Leaked documents reveal Russian effort to exert influence in Africa. *The Guardian*, 11 June. Available from: www.theguardian.com/world/2019/ jun/11/leaked-documents-reveal-russian-effort-to-exert-influence-in-africa [Accessed 23 January 2022].

Huon, P. and Ostrovsky, S., 2018. Russia, The New Power in Central Africa. *CodaStory*, 19 December. Available from: www.codastory.com/disinformation/russia-new-power-central-africa/ [Accessed 24 January 2022].

International Monetary Fund, 2020. *Direction of Trade Statistics 2020.* New York: International Monetary Fund. Available from: https://data.imf.org/regular.aspx?key=61013712 [Accessed 3 January 2022].

Limonier K., 2019. *The Dissemination of Russian-sourced News in Africa*. Paris: Institut de Recherche Strategique de l'Ecole Militaire. Available from: www.irsem.fr/data/files/irsem/documents/document/file/2965/RP_IRSEM_No66_2019.pdf [Accessed 23 January 2022].

Mackinnon A., 2021. Russia's Wagner Group Doesn't Actually Exist. *Foreign Policy*, 6 July. Available from: https://foreignpolicy.com/2021/07/06/what-is-wagner-group-russia-mercenaries-military-contractor/ [Accessed 24 January 2022].

Ministry of Digital Development, Communications and Mass Media of the Russian Federation, 2019. *Russia and Congo Agreed on Cooperation in the Field of Mass Communications*. Moscow: Ministry of Digital Development, Communications and Mass Media of the Russian Federation. Available from: https://digital.gov.ru/ru/events/39062/ [Accessed 16 January 2022].

Ministry of Foreign Affairs of the Russian Federation, 2016. *Foreign Policy Concept of the Russian Federation*. Moscow: Ministry of Foreign Affairs of the Russian Federation. Available from: www.rusemb.org.uk/in1/ [Accessed 5 January 2022].

Ministry of Information, 2019. *Eritrean Television and Russia Today (RT) Sign Cooperation Agreement*. Asmara: Ministry of Information. Available from: https://shabait.com/2019/05/15/eritrean-television-and-russia-today-rt-sign-cooperation-agreement/ [Accessed 16 January 2022].

Myers, S. L. and Bradsher, K., 2021. China's Leader Wants a "Lovable" Country. That Doesn't Mean He's Making Nice. Diplomacy. *The New York Times*, 8 June. Available from: www.nytimes.com/2021/06/08/world/asia/china-diplomacy.html.

Office of the President of the Russian Federation, 2021. *On the National Security Strategy of the Russian* Federation. Moscow: Office of the President of the Russian Federation. Available from: www.researchgate.net/publication/352947309_National_Security_Strategy_of_the_Russian_Federation_2021 [Accessed 5 January 2022].

Patrick, P., 2018. Centafrique: La Mission Russe Procède à la Remise de la Trampoline pour Enfant au Centre de la Mére et de l'Enfant de Bangui. *Le Potential Centafricaine*, 12 August. Available from: https://lepotentielcentrafricain.com/centrafrique-la-mission-russe-procede-a-la-remise-de-la-trampoline-pour-enfant-au-centre-de-la-mere-et-de-lenfant-de-bangui/ [Accessed 24 January 2022].Prinsloo, L., 2020. China's Huawei Prospers in Africa Even as Europe and Asia Join Trump's Ban. *Bloomberg Business Week*, 19 August. Available from: www.bloomberg.com/news/articles/2020-08-19/china-s-huawei-prospers-in-africa-even-as-europe-asia-join-trump-s-ban [Accessed 20 March 2022].

Proekt, 2019. Master and Chef: How Russia Interfered in Elections in Twenty Countries. *Proekt*, 11 April. Available from: www.proekt.media/en/article-en/russia-african-elections/ [Accessed 24 January 2022].

Putin, P., 2019. *Interview with TASS News Agency*. Moscow: TASS. Available from: http://en.kremlin.ru/events/president/news/61858 [Accessed 3 January 2022].

Reporters Without Borders, 2018. *China's Pursuit of a New World Media Order*. Available from: https://rsf.org/sites/default/files/en_rapport_chine_web_final.pdf [Accessed 26 March 2022].

RIA Fan, 2021. The authorities of the Central African Republic highly appreciated the film "Tourist" about the help of Russian instructors to the Republic. [Google translation]. *RIA Fan*, 16 May. Available from: https://riafan.ru/1446241-vlasti-car-vysoko-ocenili-film-turist-o-pomoshi-rossiiskikh-instruktorov-respublike [Accessed 24 January 2022].

RIA Novosti, 2019. Sputnik Expands its Presence in Africa. *RIA Novosti*, 23 May. Available from: https://ria.ru/20190523/1554844447.html [Accessed 16 January 2022].

RT, n.d. *Where to Watch*. Moscow: Russia Today. Available from: www.rt.com/where-to-watch/ [Accessed 17 January 2022].

Sauer, P., 2021. New Movie Depicting Heroic Russian Instructors in Central African Republic Linked to "Putin's Chef." *Moscow Times*, 21 May. Available from: www.the moscowtimes.com/2021/05/21/new-movie-depicting-heroic-russian-instructors-in-cent ral-african-republic-linked-to-putins-chef-a73973 [Accessed 24 January 2022].

Simoyan, M., 2012. There Is No Objectivity, Interview with Margarita Simoyan. *Daily Kommersant*, 7 April. Available from: www.kommersant.ru/doc/1911336 [Accessed 23 January 2022].

Tanchum, M., 2021. China's New Military Base in Africa: What It Means for Europe and America. *European Council on Foreign Relations*, December. Available from: https://ecfr.eu/article/chinas-new-military-base-in-africa-what-it-means-for-europe-and-amer ica/ [Accessed 22 March 2022].

The Tourist, 2021. Film. Bangui: Tourist Film Group. Available from: https://turistfilm.ru/ [Accessed 24 January 2022].

United Nations Central African Republic Sanctions Committee, 2017. *Exemptions to the Measures*. New York: United Nations. Available from: www.un.org/securitycouncil/sanctions/2127/exemptions_measures/arms-embargo [Accessed 24 January 2022].

United Nations Commission on Trade and Development, 2021. *World Investment Report Annex Table 01: FDI Inflows, by Region and Economy 1990–2020*. New York: United Nations Commission on Trade and Development. Available from: https://unctad.org/topic/investment/world-investment-report?tab=Annex%20Table [Accessed 3 January 2021].

United Nations Office of the High Commissioner on Human Rights, 2021. *Mandate of the Working Group on the Use of Mercenaries as a Means of Violating Human Rights and Impeding the Exercise of the Right of Peoples to Self-determination; the Working Group on the Issue of Human Rights and Transnational Corporations and Other Business Enterprises; the Working Group on Enforced or Involuntary Disappearances; the Special Rapporteur on Extrajudicial, Summary or Arbitrary Executions; and the Special Rapporteur on Torture and other Cruel, Inhuman or Degrading Treatment or Punishment.* New York: United Nations. Available from: https://spcommreports.ohchr.org/TMResu ltsBase/DownLoadPublicCommunicationFile?gId=26305 [Accessed 24 January 2022].

7 Competing for Influence

Authoritarian Powers in the Cyber Domain in Latin America

Fabiana Sofia Perera

Introduction

Internet access rates in Latin America are higher than for other lower- and middle-income countries. While the relatively high rates of internet connectivity could be good news for the region, the rapid pace at which technology is expanding is outpacing the state's capacity to create legal and regulatory frameworks for cyberspace and attract and cultivate online knowledge and skills. The World Bank estimates that about two out of every three people living in Latin America use the Internet, though there is wide variation within the region. For example, Haiti and Nicaragua, two of the lowest-income countries, have only 28% and 32% of their respective populations online, whereas Costa Rica's 81% internet penetration rate exceeds that of some states in the United States. About 70% of the region has a Facebook account, and three countries in the region (Argentina, Brazil, and Mexico) crack the top 20 in the world for most Twitter users. Latin America has also made impressive achievements in e-government: more than half of the countries in the region (58%, 19 countries) are above the global UN Electronic Governance Development Index (CEPAL 2022).

The region has also made progress in drafting cybercrime legislation. Worldwide, around 80% of countries have enacted cybercrime legislation (UNCTAD 2021). In the Americas (including the United States and Canada) the number is comparable: around 86% of countries have enacted legislation. Only Bolivia, Guyana, Haiti, and Suriname have failed to draft any legislation to address this. Not coincidentally, the adoption of cybercrime legislation has closely matched levels of state capacity in the region. Chile and Mexico, the only two countries in the region that are members of the Organization for Economic Co-Operation and Development (OECD), adopted legislation over a decade ago (Chile in 1993, Mexico in 2012). Countries at the lower end of state capacity have moved more slowly. An additional challenge for countries in Latin America has been the application of the legislative frameworks they have developed. As a result, though cybercrimes are being reported, rates of impunity are very high.

Though figures on impunity in cybercrimes specifically are not available, rates of impunity for other crimes highlight the challenges that law enforcement faces in Latin America in seeking to apply existing laws and regulations. The Global

DOI: 10.4324/9781003425304-8

Impunity Index concludes that no country in the Americas presents a "low degree" of convictions (Antonio le Clercq Ortega and Rodríguez Sánchez Lara 2020). Crime in Mexico, for example, is met with a great deal of impunity, with reports of up to 94.8% of reported crimes never being solved (Reina 2021). Chile, for example, reported over 16,000 cybercrimes (as defined by their own laws) in 2020 (UNSSC 2022). Apprehensions of cybercriminals are so rare that they receive wide media coverage, such as the case of a policeman from the border city of Nogales who was apprehended in 2020 and charged with extortion (Escobar 2020). Similarly, Argentinean newspapers and news portals have reported on the apprehensions of hackers, scammers, and people downloading child pornography (Ámbito 2021; El Ciudadano 2021; Análisis Digital 2021). News reports of such apprehensions highlight that these crimes are likely rare compared to other types of crimes.

Latin America as a Place with Vulnerabilities in Governance

The vulnerabilities that exist in the cyber domain in Latin America parallel other weaknesses in the region. Governance indicators point to the region's limitations with respect to government effectiveness, controlling corruption, and guaranteeing the rule of law (WGI-Interactive Data Access 2022). In cybersecurity, the region's biggest weakness is its poor governance. Governance refers to governing with and through sets of "formal and informal institutional linkages between governmental and other actors structured around shared interests in public policymaking and implementation" (Rhodes 2007; Rhodes 1997). Good governance is "epitomized by predictable, open and enlightened policymaking; a bureaucracy imbued with a professional ethos; an executive arm of government accountable for its actions; and a strong civil society participating in public affairs; and all behaving under the rule of law" (World Bank 1994).

Given the difficulty of measuring a concept like "enlightened policymaking," good governance is often taken to mean some combination of accountability, participation, inclusiveness, and transparency (Brechenmacher and Carothers 2014). Latin American countries often lag in these indicators. The majority of citizens in Latin America believe that their country's government is run by private interests (Pring and Vrushi 2019). A high proportion of citizens also report diminished trust in government as a result of their perception of the degree of corruption in the country. More alarmingly, only 39% of citizens in the region believe (their government is doing a good job-fighting corruption (Pring and Vrushi 2019). Note that perception of corruption is not a perfect measure of corruption or transparency. Indeed, corruption itself is nearly impossible to measure, and further, the link between corruption and transparency is not direct and has been contested (OECD 2021; Escaleras et al. 2010).

High levels of perceptions of corruption and low trust in the government to address it come even after the region has investigated several high-profile corruption cases in recent years. In the past few years, some prominent figures, including former Brazilian presidents Fernando Collor de Mello, Luiz Inácio "Lula" da Silva, and Michel Temer and have been found guilty of corruption and jailed (Cook 2020). Despite this conviction, the Brazilian president was re-elected in 2022. In Guatemala, former vice-president Roxana Baldetti was found guilty of fraud and corruption surrounding a project to clean up a contaminated lake (Lakhani 2018). Ecuador convicted its former president, Rafael Correa, for taking bribes in exchange for public contracts (León Cabrera 2020).

These well-publicized convictions related to corruption suggest that the region is making some progress on accountability, a second hallmark of good govern-ance (Casas-Zamora and Carter 2017). About half the countries in the region, for example, have signed on to the Open Government Partnership, an international ini-tiative aimed at increasing transparency and accountability in government. Despite these initiatives, corruption and low accountability persist: Brazil co-chaired this initiative during the presidency of Dilma Rousseff, an official who was later impeached on corruption charges (Michener and Pereira 2011).

The region also lacks the third pillar of good governance: inclusiveness. The UN Economic Commission for Latin America and the Caribbean approved a Regional Agenda for Inclusive Social Development in 2019. In publicizing the document, the executive commissioner for the organization, explained simply, "inequality defines our region" (ECLAC 2020). In Latin America (including the Caribbean), the poorest 50% of the population earns just 10% of total income, while the wealthiest 10% controls 55% (World Bank 2022). Countries in the region suffer from economic and social inequality marked by differences in race and eth-nicity and cleavages between rural and urban populations. Though a number of governments in the region tried to address this inequality during the so-called left-turn in Latin America, the efforts to do so where themselves of varying scope and quality, which rendered mixed results (Balán and Montambeault 2020).

Finally, the region performs better on participation, the fourth pillar of good governance. This is partly due to the electoral systems in place in more than half of the countries which include compulsory voting. The actual picture of participa-tion might be more complex than the voter turnout numbers suggest; indigenous groups and other minorities, as well as women, have lower levels of participation in the region than other groups. Moreover, the causal relationship between partici-patory institutions as a dimension of good governance and quality decision-making remains unclear (Montambeault 2016; Silva and Cleuren 2009).

Weak Governance Creates Opportunities for Authoritarian Powers

As the preceding section explained, Latin America as a region is struggling to achieve good governance. The weaknesses it has in this area – high levels of corruption, high inequality and feelings to exclusion, and low accountability – are not specific to one issue area, but rather exist across most areas of governance, including governance of cyberspace and critical infrastructure. These vulnerabil-ities in turn create opportunities for authoritarian states to increase their presence in Latin America. The connection between weaknesses in governance (and on corruption specifically) and opportunities for near-peer competitors has been highlighted by the Commanders of U.S. Southern Command, the U.S. regional combatant command whose area of responsibility includes all of South America, Central America except Mexico, and nearly every Caribbean state.

The most recent (2022) posture statement by General Laura Richardson, commander of U.S. Southern Command, identified a "trail of corruption and violence that create[s] conditions that allow the PRC and Russia to exploit, threaten citizen security, and undermine public confidence in government institutions" (Gen. Richardson 2022). In 2020, Admiral Craig S. Faller accused out the People's Republic of China and "other external state actors" of exploiting the conditions created by the threat of corruption, among others (ADM Faller 2021). The previous year, ADM Faller highlighted "weak governance and porous legal frameworks" in the region as conditions favorable to China and Russia (ADM Faller 2020). Tellingly, similar documents from other regions do not highlight poor governance as a vulnerability for near-peers to exploit, though some of them do point to engagement by the PRC and Russia as contributing to weak governance (ADM Davidson 2021).

What Goals Are Competitors Pursuing?

Authoritarian competitors challenge the free and open international order and wish to shape international norms to obtain diplomatic, economic, and strategic advantages in Latin America. The primary competitors to the United States in the region, China and Russia, are examples of "economically successful authoritarian capitalist powers" (Gat 2007). Differences in ideology are a primary difference between the democratic United States and its authoritarian challengers. The highest level of leadership in the United States referred to the competition between the United States and its challengers as a "battle between democracy and autocracies" (Biden 2022). These differences not only motivate competition but they also shape the strategies they use in this competition (Brands 2014). While the United States engages in the name of making the world safe for democracy, authoritarian powers seek to break perceived American ideological hegemony. In Russia, President Vladimir Putin opposes a world "in which there is only one master, one sovereign" (Fried and Volker 2022). China, meanwhile, has sought to redefine democracy away from a "Western-centric yardstick" and toward a process in which "the people are the masters of the country" (Saha 2022). More forcefully, President Xi Jinping has warned against attempts to "bully" the country (Lun Tian and Woo 2021).

Though both Russia and China seek to challenge American hegemony and are actively engaged around the world in this pursuit, each competitor is pursuing different goals in Latin America. Under the leadership of Putin, Russia has shown a desire to build on its long history of regional engagement to strengthen relations with Latin American and Caribbean states. The twin goals of the reengagement are first, to boost Russia's credentials as a global power, and second, to maintain a presence in the United States' near abroad as a check against the United States' involvement in Russia's near abroad (e.g., Georgia, Ukraine, the Baltic states). Russia was able to call on this recently when it threatened to position military assets in Cuba and Venezuela should the United States further engage in Ukraine (Cowen 2022). Chinese ties to the region, on the other hand, are much shallower: their longest diplomatic relationships with Latin America, those with Mexico and Argentina, will be just 50 years old next year. Chinese interest in the region is motivated by a desire to secure access to natural resources and markets (Da Rocha and Beilschowsky 2018).

Cyber Security in Latin America

The cyber domain invites the use of all instruments of power: diplomatic, information, military, and economic (DIME). Indeed, the extensive reach of the cyber domain and the actions that are possible within it could make the DIME mnemonic obsolete as a summary of the instruments of national power (Thomas 2014). Whereas the cyber domain is sometimes referred to as the "information domain" states are also able to use it for military and economic purposes. The United States Department of Defense Cyber Strategy specifically acknowledges that adversaries are operating in cyberspace to "erode U.S. military advantages, threaten our infrastructure, and reduce our economic prosperity," actions that are clearly outside the information space (Department of Defense 2018). In this context, cybersecurity refers to efforts to prevent systems from being compromised (Libicki 2016).

Both Russia and China have been pragmatic in their approach to the region, displaying flexibility and an ability to adapt quickly as they pursue economic and security objectives (Giusti and Penkova 2008; Piccone 2020). The two authoritarian states have sought to employ cyber tools including disinformation, cybercrime, electoral interference, and investment in critical infrastructure in different ways as part of their strategy for the region. Broadly, Russia has placed greater emphasis on the use of cyber tools as a military and information tool. This is consistent with their focus on diplomatic and military engagement with the region. China, in contrast, has mostly leveraged cyber as an economic instrument of power.

Russia and Cyber in Latin America

Consistent with establishing a presence close to the United States, Russia has used cyber to expand its online presence in Latin America, as well as in support of its military and diplomatic objectives and engagements. Russia has most significantly expanded its use of information tools in the cyber domain through the establishment and expansion of RT en Español, part of its state-controlled propaganda network (RT en Español 2022a,b). RT en Español began its broadcast operations in 2009 and has since expanded across platforms to include online streaming through social media platforms. According to RT, its Facebook channel in Spanish has more followers than CNN (RT en Español 2022a,b). Spanish-speaking audiences are an important target for RT, accounting for about of all RT videos made available through YouTube 11% (Orttung and Nelson 2019). Spanish is one of the top priority languages for RT after Arabic, Russian, and English (Orttung and Nelson 2019). Worldwide, RT en Español has 205 million followers, nearly twice as many as El Mundo, one of Spain's newspapers of record (Rebello et al. 2020).

RT and the news agency Sputnik are critical elements in Russia's propaganda campaigns. In February 2022, the European Union banned the two outlets clarifying that the outlets are not examples of independent media but rather "assets" and "weapons, in the Kremlin's manipulation ecosystem" (Petrequin 2022). Putin himself has said that RT is meant to "reflect the Russian government's official position on the events in our country and in the rest of the world, one way or another," a vision that presumably extends to its online outlets. As an example of this, recent headlines on the site have included statements like "The White House is not opposed to Kyev attacking Crimea" and "The White House wants to avoid World War 3, but it sends weapons to Ukraine" (RT en Español 2023a,b).

Russia has used online propaganda both to create a narrative about its position in the world and to disrupt domestic events in Latin America. The first use is obvious in the examples above and in its reporting on the Russian invasion of Ukraine, which includes headlines about increases in the price of food and reports of a Venezuelan official denouncing that over 30 countries are "victims of coercive measures taken by the West." It also included instructions on how to circumvent YouTube's blocking of RT content and a report that Facebook's parent company, Meta, had temporarily changed its hate speech policy to allow messages of hate directed at Russian soldiers. RT has also produced online reports about Russia's leading role in vaccine diplomacy in Latin America (e.g., "Sputnik is the most common vaccine in Mexico City" and "Several Latin American countries opted in 2021 for the Russian vaccine Sputnik V" (RT en Español 2022a,b).

Russia has also used its news outlets and their accompanying websites and social media sites to disrupt domestic politics in Latin America. Ahead of the Mexican presidential elections in 2018, RT en Español began broadcasting "The Battle for Mexico," a weekly video blog hosted by an American-born, naturalized-Mexican professor and political activist. The host, John Ackerman, was a strong supporter of the eventual winner, Andrés Manuel López Obrador. The host portrayed López Obrador as the only candidate willing to stand up to President Donald Trump (Salvo and de Leon 2018). HR McMaster, the former U.S. National Security Adviser, denounced "signs of [Russian] interference" in the 2018 election, highlighting that there is concern that Russia is using "cyber tools" in "sophisticated campaigns of subversion and disinformation and propaganda" (Shelbourne 2018).

In addition to these information and propaganda campaigns, Russia has also engaged in cyberspace in support of its diplomatic and military objectives to directly interfere with countries in the region. These tools have included the use of malware, ransomware, and Trojan attacks. For example, in September 2020, the Federation of American Scientists (FAS) reported that Sputnik's Spanish-language website was at the center of a network of websites hosting malware files. The Federation's analysis found malware associated with popular news sites covering Oxford-AstraZeneca-related stories. The discovery was made by scanning URLs covering the Oxford-AstraZeneca vaccine. The scans returned 53 sites hosting malware within the AstraZeneca conversation (Disinformation Research Group 2020). The placement of malware in these sites may provide opportunities for perpetrators to manipulate additional web traffic to unwitting users to inorganically amplify stories that could cast doubt on the efficacy of particular vaccines. These malware files could be related to a broader Russian campaign to create distrust around Western-backed vaccines. The FAS hypothesizes that the malware is used to identify and track an audience interested in the state of COVID-19 vaccines so that Russia can target this specific group to create more favorable views of its own vaccine (Disinformation Research Group 2020).

The region has also been the victim of other malware attacks, including a "massive rise in trojans," especially in the financial sector. Russian-supported hackers have also been blamed for ransomware attacks on region (O'Brien 2015). In 2019, Latin America was hit with Ryuk ransomware attacks. One attack completely shut down PEMEX, Mexico's state-owned oil company. Ryuk is especially dangerous because it was created to infect a system and remain undetected for a period of time, during which the malware seeks out critical network systems to maximize its impact (Constantin 2021). Though it is not possible to know its true origins, cultural references made during negotiations and other data have led researchers to link the malware to Russia (Fokker 2019).

In addition to the attacks linked to propaganda and ransomware attacks targeting critical infrastructure, Russia has also been accused of using malware to interfere with elections in the region. In the run-up to its parliamentary election in 2018, Colombia detected more than 50,000 attacks on the web platform of its national voter registry. The country's minister of defense said the attacks were staged in Venezuela, and another official added that "Russia is increasingly using Venezuela as a base for covert operations in its growing rivalry with the U.S. for international influence" (Arostegui 2018).

China and Cyber in Latin America

In contrast to Russia, China has not pursued operations in the information environment in Latin America, even though it does employ soft power and cultural diplomacy tools extensively. China's Confucius Institutes most famously embody soft power tools. These are Chinese language and cultural centers funded by grants from China. In the United States, they were designated as "foreign missions" by the State Department and are subject to close monitoring (Riechmann 2020). In Latin America, China has established forty-three Confucius Institutes to date, an impressive rate of growth considering that the first institute opened in 2006 (Gilstrap 2021). China is standing up institutes as quickly as it can: Panama recognized China in June 2017 and almost exactly a year later, the first Confucius Institute opened in its capital (Xinhua Español 2018).

China's efforts to create goodwill in Latin America have focused not on propaganda campaigns using new websites and outlets but on working with existing news outlets and journalists. In its efforts to create links between regional media and its own news agencies, China has appeared to emphasize print outlets, inviting journalists to visit China (Cardenal 2017). Given this strategy, its efforts to influence the online information environment have so far paled in comparison to its more robust cultural and economic engagement and most definitely have not been as strong or aggressive as Russia's. Though the PRC's news agency, Xinhua, does have a Spanish-language website, its presence is not as robust as that of RT en Español.

In the cyber domain, China has primarily sought to engage through its strongest links to Latin America, the private sector. This is consistent with the PRC's persistent focus on economic tools in pursuit of its goals. Chinese firms have been very active in two areas that could impact the cyber domain: providing telecommunications networks and providing security technologies. Chinese telecommunication firms have proactively attempted to influence international standards governing 5G networks. In this regard, China's efforts in Latin America have paid off. Huawei, for example, has reported an increase in revenue driven by its Latin American operations (Labs 2020). In the region, China has had an important role in developing telecommunication infrastructure. In Brazil, Huawei created most of the 3G and 4G infrastructure and will probably compete to build its 5G network (Silva Ramos Becard and Vieira De Macedo 2014; Reuters 2021). In Mexico, Huawei won the contract to supply equipment for telecommunications upgrades (Love 2019). Huawei will also be involved in Venezuela through a project to expand the country's communications network (Reimi 2018).

In advancing itself as a provider of 5G technology, China is seeking to dominate this space. Unlike in the United States, where many communications technology sectors tend to run "parallel but separate," China is seeking to intertwine everything (Arrington 2022). Though Huawei, the main possible PRC provider of 5G technology, is not a state-owned company, China's national security laws require individuals and companies (including "employee-owned" Huawei) to cooperate with PRC security services, which could turn any Huawei-provided technology into a potential threat to privacy. Of particular concern to the United States is that through Huawei, Beijing could spy on or disrupt infrastructure and operations (Atlamazoglou 2021). Advanced technologies that rely on 5G networks, including autonomous vehicles, drones, robotics, augmented reality/virtual reality (AR/VR), and video surveillance cameras, could all prove vulnerable (Arrington 2022).

In addition to its push to become a provider of telecommunications networks and infrastructure in the region, China is also exporting its security technology. Recently, Chinese companies like DuHua started leveraging the COVID-19 pandemic to expand their footprints in the region. DuHua donated thermal imaging cameras to Argentina, Chile, Colombia, Mexico, and Panama (Gerencia 2023; Tecno Seguro 2020; Revista Mas Seguridad 2022; León Barría 2020). Though these firms are not state-owned, they do receive significant financial backing from China and act as state proxies. Already in 2017, Venezuela started working with ZTE, another Chinese communications giant, helped create the regime's "fatherland card" (Berwick 2018). The card is needed to buy subsidized food and gas, prompting concerns that it could be used to allocate much-needed resources primarily to supporters of the regime. Regarding these allegations, Su Qingfeng, the head of ZTE's Venezuela unit, said, "We don't support the government. We are just developing our market," which is consistent with the language that China has used in discussing its engagement in Latin America (Berwick 2018). Despite repeating its stated aims of "market expansion" and economic opportunity, China's approach to citizen security through its "safe cities" program has been accused by the United States of "exporting authoritarianism," an effect that is in direct opposition to the U.S. objectives (Hillman and McCalpin 2019; ADM Faller 2021) of supporting a safe and secure Western Hemisphere bound by democratic principles and values.

China's engagement with Latin America in the cyber domain has been consistent with its broader strategy in the region. It has prioritized or claimed to prioritize economic opportunity for itself and for its regional partners. In cyber, much of this economic opportunity is represented by access to technologies that, while ostensibly sold for civilian use or civilian protection, have a potential dual use that could give China a strategic advantage in the region.

In other words, foreign investment in infrastructure is not, in and of itself, a security concern for Latin America. The Pan-American Highway, for example, was partly financed by the United States in some countries. In the case of Chinese investments in this sector, however, the concern is that the technologies they are bringing to the region, as well as some of the ones they are developing at home (e.g., quantum computing, AI), are dual-use technologies – that is, technologies that through their normal civilian use also give China and its partners a military advantage in the region.

Conclusion

Authoritarian powers are attracted to Latin America because the region constitutes the United States' near-abroad even if the U.S. seldom engages with it this way, choosing instead to refer to its hemisphere, its neighborhood, or, in less-enlightened times, its backyard. The region's deficiencies in governance create openings for authoritarian powers to engage. Taking advantage of high degrees of corruption and impunity, U.S. near-peers who are not committed to the ideals of democracy and transparency can seek to turn these countries away from the United States and toward new partnerships that, while flaunted as "no-strings attached" come with lots of "strings." It's just different strings.

Authoritarian powers have strengthened their position in the region through the use of cyber tools that have included disinformation campaigns, provision of telecommunications technology, and cyberattacks. Though great power competition is often discussed with a focus on the motivations and agency of the great powers themselves, it's foolish to consider Latin American countries pawns devoid of agency. Latin American countries have agency and do choose with whom to engage, for what, and when. The challenge is that poor governance directly impacts the agency of the countries, hampering their ability to make decisions that are in their best interest long-term.

Strengthening governance in the region can help countries evaluate foreign offers of investment and determine what might be in their own country's long-term interests. As a top Brazilian official explained at a recent cybersecurity conference focused on Latin America, "good national governance entails top-down strategic planning" to focus on what is relevant and best for the country (Inter-American Defense Foundation 2020). Working with partner nations to strengthen capacity presents a promising avenue for the United States to counter Russian and Chinese advancements in the cyber domain. Additionally, as the chairman of the Inter-American Defense Board, General Luciano Penna, explained at that same conference, cyber defense "represents the best opportunity for cooperation in the Western Hemisphere." The United States support to Latin American countries as they work collaboratively to improve cyber governance in the region could result in a diminished presence of competitors in the cyber domain.

References

Ámbito, 2021. Detienen a Joven por Estafa Millonaria a Mercado Libre [A Young Man Is Arrested for Millionaire Fraud to Mercado Libre]. *Ámbito*, 28 November. Available from: www.ambito.com/informacion-general/mercado-libre/detienen-joven-estafa-mil lonaria-n5325509

Análisis Digital, 2021. Gualeguaychú: Detienen a Un Hombre en Un Ciber por Descargar Pornografía Infantil [Gualeguaychú: A Man Is Arrested in a Cyber for Downloading Child Pornography]. *Análisis Digital*, 14 August. Available from: www.analisisdigital.com.ar/policiales/2021/08/14/gualeguaychu-detienen-un-hombre-en-un-ciber-por-descargar-pornografia-infantil

Antonio le Clercq Ortega, J. and Rodríguez Sánchez Lara, G., 2020. *Escalas de Impunidad en el Mundo* [Scales of Impunity in the World]. San Andrés Cholula, Puebla, Mexico: Universidad de las Américas Puebla. Available from: www.udlap.mx/cesij/files/indices-globales/0-IGI-2020-UDLAP.pdf

Arostegui, M., 2018. Colombia Probes Voter Registration Cyberattacks Traced to Russia's Allies. *VOA News*, 15 March. Available from: www.voanews.com/a/colombia-voter-regis tration-cyberattacks-russia-allies/4300571.html

Arrington, G., 2022. It's Not Just 5G: China's Telecom Strategy Needs to be Countered in Space. *Breaking Defense*, 14 February. Available from: https://breakingdefense.com/2022/02/its-not-just-5g-chinas-telecom-strategy-needs-to-be-countered-in-space/

Atlamazoglou, S., 2021. Special Operators Are Already Dealing with a Shady Piece of Chinese Technology the US Has Been Warning About. *Business Insider*, 31 August. Available from: www.businessinsider.com/us-military-personnel-already-face-risks-from-chinese-5g-technology-2021-8

Balán, M. and Montambeault, F., 2020. *Legacies of the Left Turn in Latin America: The Promise of Inclusive Citizenship*. Notre Dame, Indiana: University of Notre Dame Press. Available from: https://books.google.com/books?hl=es&lr=&id=vyvJDwAAQBAJ&oi=fnd&pg=PT10&dq=Balán+and+Montambeault,+2020&ots=k0Slxzk8CG&sig=7gr Cw3CBmFsiea-qXnNPRuTmVQE#v=onepage&q=Balán%20and%20Montambea ult%2C%202020&f=false

Berwick, A., 2018. How ZTE Helps Venezuela Create China-Style Social Control. *Reuters*, 14 November. Available from: www.reuters.com/investigates/special-report/venezu ela-zte/

Biden, J. R., 2022. Remarks by President Biden on the United Efforts of the Free World to Support the People of Ukraine. 26 March, Warsaw, Poland. Available from: www.whi tehouse.gov/briefing-room/speeches-remarks/2022/03/26/remarks-by-president-biden-on-the-united-efforts-of-the-free-world-to-support-the-people-of-ukraine/

Brands, H., 2014. *What Good Is Grand Strategy? Power and Purpose in American Statecraft from Harry S. Truman to George W. Bush*. Ithaca, NY: Cornell University Press.

Cardenal, J.P., 2017. Sharp Power: Rising Authoritarian Influence. In: *National Endowment for Democracy*, 26–36. Available from: www.ned.org/wp-content/uploads/2017/12/Chapter1-Sharp-Power-Rising-Authoritarian-Influence-China-Latin-America.pdf

Carothers, T. and Brechenmacher, S., 2014. Accountability, Transparency, Participation and Inclusion: A New Development Consensus? *Carnegie Endowment for International Peace*. Available from: www.jstor.org/stable/resrep12957

Casas-Zamora, K. and Carter, M., 2017. *Beyond the Scandals: The Changing Context of Corruption in Latin America*. Washington D.C.: Inter-American Dialogue. Available from: www.thedialogue.org/wp-content/uploads/2017/02/Corruption-in-Latin-America_ROL_Report_FINAL_web-PDF.pdf

CEPAL, 2022. Latin American and the Caribbean Countries are Highly Committed to Pursuing Digital Government Strategies, but Inclusivity and E-participation Remain a Challenge. *CEPAL*, 19 December. Available from: www.cepal.org/en/news/latin-american-and-caribbean-countries-are-highly-committed-pursuing-digital-government

Constantin, L., 2021. Ryuk Explained: Targeted, Devastatingly Effective Ransomware. *CSO Online*, 19 March. Available from: www.csoonline.com/article/3541810/ryuk-explained-targeted-devastatingly-effective-ransomware.html

Cook, 2020. Lessons from Brazil's Fight against Corruption. *Diálogo Américas*, 11 March. Available from: https://dialogo-americas.com/articles/lessons-from-brazils-fight-against-corruption/#.ZFukjS_MJeh

Cowen, C., 2022. How Russia Militarizes Authoritarianism in Latin America. *National Interest*, 29 July. Available from: https://nationalinterest.org/blog/buzz/how-russia-militarizes-authoritarianism-latin-america-203866

Da Rocha, F.F. and Beilschowsky, R., 2018. China's Quest for Natural Resources in Latin America. *Cepal Review*, 126, 9–28. Available from: www.cepal.org/en/publications/44555-chinas-quest-natural-resources-latin-america

Davidson (ADM), P. S., 2021. Statement of Admiral Philip S. Davidson, U.S. Navy Commander, U.S. Indo-Pacific Command. *Senate Armed Services Committee*, 9 March. Available from: www.armed-services.senate.gov/imo/media/doc/Davidson_03-09-21.pdf

Department of Defense, 2018. Summary: Department of Defense Cyber Strategy. *United States Department of Defense*. Available from: https://dodcio.defense.gov/Portals/0/Documents/Library/CyberStrategy2018.pdf

Disinformation Research Group, 2020. Vaccine News Stories Hosting Malware Disseminated Across Spanish-Language Twitter. *FAS*. Available from: https://fas.org/publication/vaccine-news-stories-hosting-malware-disseminated-across-spanish-language-twitter/

Economic Commission for Latin America and the Caribbean (ECLAC), 2020. Regional Agenda for Inclusive Social Development (LC/CDS.3/5). Santiago: United Nations ECLAC. Available from: www.cepal.org/en/publications/45330-regional-agenda-inclusive-social-development

El Ciudadano, 2021. Detienen a Cinco Integrantes de Una Banda Acusada de Hackear una Cuenta Bancaria y Desviar Fondos [Five Members of a Gang Accused of Hacking a Bank Account and Diverting Funds Arrested]. *El Ciudadano*, 16 August. Available from: www.elciudadanoweb.com/detienen-a-cinco-integrantes-de-una-banda-acusada-de-hackear-una-cuenta-bancaria-y-desviar-fondos/

Escaleras, M., Lin, S. and Register, C., 2010. Freedom of Information Acts and Public Sector Corruption. *Public Choice*, 145, 435–460. https://doi.org/10.1007/s11127-009-9574-0

Escobar, A., 2020. Detienen a Policía Acusado de Extorsión en Sonora; Intentó Irse a EU [Police Officer Accused of Extortion Arrested in Sonora; He Tried to Flee to the US]. *El Universal*, 20 February. Available from: www.eluniversal.com.mx/estados/detienen-poli cia-acusado-de-extorsion-en-sonora-intento-irse-eu/

Faller (ADM), C.S., 2020. Posture Statement of Admiral Craig S. Faller Commander, United States Southern Command. *United States Southern Command*, 30 January. Available from: www.southcom.mil/Portals/7/Documents/Posture%20Statements/ SASC%20SOUTHCOM%20Posture%20Statement_FINAL.pdf?ver=2020-01-30-081 357-560

Faller (ADM), C. S., 2021. Statement of Admiral Craig S. Faller Commander, United States Southern Command. *United States Southern Command*, 17 March. Available from: www. southcom.mil/Portals/7/Documents/Posture%20Statements/SOUTHCOM%202021%20 Posture%20Statement_FINAL.pdf?ver=qVZdqbYBi_-rPgtL2LzDkg%3D%3D

Fokker, J., 2019. Ryuk Ransomware Attack: Rush to Attribution Misses the Point. *McAfee Blog*, 18 January. www.mcafee.com/blogs/other-blogs/mcafee-labs/ryuk-ransomware-att ack-rush-to-attribution-misses-the-point/

Fried, D. and Volker, K., 2022. The Speech in Which Putin Told Us Who He Was. *Politico*, 2 February. Available from: www.politico.com/news/magazine/2022/02/18/putin-speech- wake-up-call-post-cold-war-order-liberal-2007-00009918

Gat, A., 2007. The Return of Authoritarian Great Powers. *Foreign Affairs*, 86, 59–69. Available from: www.foreignaffairs.com/articles/china/2007-07-01/return-authoritarian- great-powers

Gerencia, 2023. El punto neurálgico de la ciberseguridad en las empresas. *Gerencia*, 3 April. Available from: www.emb.cl/gerencia/noticia.mvc?nid=20200403w9&ni=aeropuerto- de-iquique-instala-camara-termica-de-dahua

Gilstrap, J., 2021. Chinese Confucius Institutes in Latin America: Tools of Soft Power. *National Defense University*. Available from: https://library.villanova.edu/Find/Record/ 2847220

Giusti, S. and Penkova, T., 2008. From Ideology to Pragmatism: The New Course of Russian Foreign Policy. *World Affairs: The Journal of International Issues*, 12(4), 14–53.

Hillman, J. and McCalpin, M., 2019. Watching Huawei's "Safe Cities." *Center for Strategic and International Studies*, 4 November. Available from: www.csis.org/analysis/watching- huaweis-safe-cities

Inter-American Defense Foundation, 2020. Cyber Defense Conference. *Inter-American Defense Foundation*, February. Available from: www.iadfoundation.org/wp-content/uplo ads/2021/02/ciber-ingles-_compressed-4.pdf

LABS, 2020. Even with US Pressure, Huawei Sees Revenue Grow 18% in 2019. *Latin America Business Stories*, 2 January. Available from: https://labsnews.com/en/news/tec hnology/even-with-us-pressure-huawei-sees-revenue-grow-18-in-2019/

Lakhani, N., 2018. Guatemala's Former Vice-President Jailed for 15 Years on Corruption Charges. *The Guardian*, 9 October. Available from: www.theguardian.com/world/2018/ oct/09/guatemala-former-vice-president-jailed-15-years-corruption-case

León Barría, G., 2020. Dahua Technology Panamá Dona Segunda Cámara Térmica al Minsa [Dahua Technology Panama Donates Second Thermal Camera to Minsa]. *La Estrella de Panamá*, 8 April. Available from: www.laestrella.com.pa/nacional/200408/dahua-technol ogy-panama-dona-segunda-camara-termica-minsa

León Cabrera, J.M., 2020. Ecuador's Former President Convicted on Corruption Charges. *The New York Times*, 7 April. Available from: www.nytimes.com/2020/04/07/world/ americas/ecuador-correa-corruption-verdict.html

Libicki, M., 2016. Is There a Cybersecurity Dilemma? *The Cyber Defense Review*, 1(1), 129–140.

Love, J., 2019. U.S. Campaign against Huawei Hits a Snag South of the Border. *Reuters*, 9 May. Available from: www.reuters.com/article/us-mexico-huawei-tech-insight/u-s-campaign-against-huawei-hits-a-snag-south-of-the-border-idUSKCN1SF15Z

Lun Tian, Y. and Woo, R., 2021. Xi Warns against Foreign Bullying as China Marks Party Centenary. *Reuters*, 30 June. Available from: www.reuters.com/world/china/beijing-set-celebrate-centenary-chinas-communist-party-2021-06-30/

Michener, G. and Pereira, C., 2011. Is Brazil Fit to Lead the Open Government Partnership? Secrecy vs. Transparency and the Ambivalence of Brazil's Presidents. *The Brookings Institution*, 18 July. Available from: www.brookings.edu/opinions/is-brazil-fit-to-lead-the-open-government-partnership-secrecy-vs-transparency-and-the-ambivalence-of-brazils-presidents/

Montambeault, F., 2016. *The Politics of Local Participatory Democracy in Latin America: Institutions, Actors, and Interactions*. Stanford, CA: Stanford University Press.

O'Brien, R., 2015. Latam Cyber Attacks Rise as Peru, Brazil Hackers Link Up with Russians. *Reuters*, 28 August. Available from: www.reuters.com/article/latam-cyberattack/latam-cyber-attacks-rise-as-peru-brazil-hackers-link-up-with-russians-idUKL5N1134W520150828

OECD, 2021. *Government at a Glance 2021*. Paris: OECD Publishing. Available from: www.oecd.org/gov/government-at-a-glance-22214399.htm

Orttung, R. and Nelson, E., 2019. Russia Today's Strategy and Effectiveness on YouTube. *Post-Soviet Affairs*, 35(2), 77–92. https://doi.org/10.1080/1060586X.2018.1531650

Petrequin, S., 2022. EU Pledges to Fight Russia's "Information War" in Europe. *Associated Press*, 8 March. Available from: https://apnews.com/article/russia-ukraine-business-europe-media-european-union-95c44aeb2e4182227e40ae9fbe6b5841

Piccone, T., 2020. China and Latin America: A Pragmatic Embrace. *The Brookings Institution*, July. Available from: www.brookings.edu/research/china-and-latin-america-a-pragmatic-embrace/

Pring, C. and Vrushi, J., 2019. Citizens' Views and Experiences of Corruption. *Transparency International*. Available from: www.transparency.org/en/gcb/latin-america/latin-america-and-the-caribbean-x-edition-2019

Rebello, K., Schwieter, C., Schliebs, M., Joynes-Burgess, K., Elswah, M., Bright, J., and Howard, P. N., 2020. COVID-19 News and Information from State-Backed Outlets Targeting French, German and Spanish-Speaking Social Media Users. *University of Oxford*, Comprop Data Memo 2020.4, 29 June. Available from: https://demtech.oii.ox.ac.uk/wp-content/uploads/sites/12/2020/06/Covid-19-Misinfo-Targeting-French-German-and-Spanish-Social-Media-Users-Final.pdf

Reimi, I., 2018. La Inquietante Presencia de Huawei En Venezuela [The Worrying Presence of Huawei in Venezuela]. *El Estímulo*, 15 December. Available from: https://elestimulo.com/tecnologia/2018-12-15/la-inquietante-presencia-de-huawei-en-venezuela/

Reina, E., 2021. La Impunidad Crece en México: Un 94,8% de Los Casos No Se Resuelven [Impunity Grows in Mexico: Some 94.8% of Cases Are Not Resolved]. *El País*, 5 October. Available from: https://elpais.com/mexico/2021-10-05/la-impunidad-crece-en-mexico-un-948-de-los-casos-no-se-resuelven.html#

Reuters, 2021. Brazil Regulator Approves 5G Spectrum Auction Rules, No Huawei Ban. *Reuters*, 26 February. Available from: www.reuters.com/business/media-telecom/brazil-regulator-approves-5g-spectrum-auction-rules-no-huawei-ban-2021-02-26/

Rhodes, R.A.W., 1997. *Understanding Governance: Policy Networks, Governance, Reflexivity and Accountability*. Philadelphia, PA: Open University. Available from: https://eprints.soton.ac.uk/336524/

Rhodes, R.A.W., 2007. Understanding Governance: Ten Years On', *Organizational Studies*, 28(8), 1243–1264. https://doi.org/10.1177/0170840607076586

Richardson (Gen.), L. J., 2022. Statement of General Laura J. Richardson Commander, United States Southern Command. *United States Southern Command*, 8 March. Available from: www.southcom.mil/Portals/7/Documents/Posture%20Statements/SOUTHCOM%20Posture%20Final%202022.pdf?ver=tkjkieaC2RQMhk5L9cM_3Q%3D%3D

Riechmann, D., 2020. Trump Administration: Confucius Institute Is Arm of Beijing. *The Washington Post*, 13 August. Available from: www.washingtonpost.com/world/national-security/trump-administration-confucius-institute-is-arm-of-beijing/2020/08/13/37418da0-dd8a-11ea-b4f1-25b762cdbbf4_story.html

RT en Español, 2022a. RT en Español [RT in Spanish]. *Russia Today*. Available from: https://actualidad.rt.com [banned in the EU as of 2 March 2022].

RT en Español, 2022b. Quiénes Somos [About Us]. *Russia Today*. Available from: https://actualidad.rt.com/acerca/quienes_somos [banned in the EU as of 2 March 2022].

RT en Español, 2023a. La Casa Blanca Quiere Evitar la Tercera Guerra Mundial Pero Envía Armas a Ucrania [The White House Wants to Avoid World War III But Sends Weapons to Ukraine]. *Russia Today*, 20 May. Available from: https://actualidad.rt.com/actualidad/467628-eeuu-actuar-evitar-tercera-guerra-mundial [banned in the EU as of 2 March 2022].

RT en Español, 2023b. La Casa Blanca No Se Opone a Que Kiev Ataque Crimea [The White House is Not Oppose Kyiv Attacking Crimea]. *Russia Today*, 22 May. Available from: https://actualidad.rt.com/actualidad/467758-casa-blanca-no-opone-kiev-ataque-crimea [banned in the EU as of 2 March 2022].

Saha, R., 2022. U.S.-ASEAN Summit: Democracy Promotion on the Backburner. *Institute for Security and Development Policy*, 9 September. Available from: www.isdp.eu/content/uploads/2022/09/Brief-Sept-9-2022-Rushali-final.pdf

Salvo, D. and de Leon, S., 2018. Russia's Efforts to Destabilize Bosnia and Herzegovina. *German Marshall Fund of the United States* [preprint]. Available from: www.jstor.org/stable/pdf/resrep18769.pdf

Shelbourne, M., 2018. McMaster: Russia Interfering in Mexican Election. *The Hill*, 1 July. Available from: https://thehill.com/homenews/administration/367844-mcmaster-points-to-russian-interference-in-mexican-election/

Silva, P. and Cleuren, H., 2009. *Widening Democracy: Citizens and Participatory Schemes in Brazil and Chile*. Leiden, The Netherlands: Koninklijke Brill NV. Available from: https://books.google.com/books?hl=es&lr=&id=fjmwCQAAQBAJ&oi=fnd&pg=PP7&dq=Silva+%26+Cleuren+2009&ots=0b_p8-WofM&sig=FC4ANXTVCvFV6h9HFCHqMl20WJE#v=onepage&q=Silva%20%26%20Cleuren%202009&f=false

Silva Ramos Becard, D. and Vieira de Macedo, B., 2014. Chinese Multinational Corporations in Brazil: Strategies and Implications in Energy and Telecom Sectors. *Revista Brasileira de Política Internacional*, 57(1), 143–161. http://dx.doi.org/10.1590/0034-7329201400107

Tecno Seguro, 2020. Cámara Térmica de Dahua Technology en el Centro Hospitalario Transitorio de Bogotá. *Tecno Seguro*, 19 May. Available from: www.tecnoseguro.com/noticias/cctv/camaras-termicas-dahua-technology-centro-hospitalario-transitorio

Thomas, T., 2014. Creating Cyber Strategists: Escaping the "DIME" Mnemonic. *Defence Studies*, 14(4), 370–393. https://doi.org/10.1080/14702436.2014.952522

United Nations Conference on Trade and Development (UNCTAD), 2021. Cybercrime Legislation Worldwide. *UNCTAD*. Available from: https://unctad.org/page/cybercrime-legislation-worldwide

United Nations System Staff College (UNSSC), 2022. United Nations System Staff College. *UNSSC*. Available from: www.unssc.org/sites/default/files/node/vacancy/field_vacancy_file/2022-02/AF_001_2022_0.pdf

WGI-Interactive Data Access, 2022. World Governance Indicators. *World Bank*. Available from: https://info.worldbank.org/governance/wgi/Home/Reports

World Bank, 1994. *Governance: The World Bank's Experience*. Available from: https://documents1.worldbank.org/curated/en/711471468765285964/pdf/multi0page.pdf

World Bank, 2022. *Poverty and Shared Prosperity 2022*. Available from: www.worldbank.org/en/publication/poverty-and-shared-prosperity

Xinhua Español, 2018. Abre Sus Puertas el Primer Instituto Confucio en Panamá [The First Confucius Institute in Panama Opens Its Doors]. *Xinhua Español*, 17 June. Available from: http://spanish.xinhuanet.com/2018-06/17/c_137260350.htm

8 The Logic of Protraction in Cyber Conflict

Peace Would Ruin Me

Trey Herr, Emma Schroeder, and Stewart Scott

Introduction – A Stable Insecurity

Escaping the daily deluge of cyber-incident headlines is nearly impossible – from log4Shell to SolarWinds, NOBELIUM to HAFNIUM, and all the pipeline shutdowns and shuttered government websites in between, the cyber domain appears in perpetual tumult. Settling on any universal measure of that cyber tumult is difficult, but most metrics paint a picture as dire as the headlines. Sonatype (2021) found that potent software supply chain attacks on open-source code increased by 650% in 2021. An Accenture report (Dal Cin et al. 2021) found that cyberattacks on corporations increased by 31% in 2021. McAfee and CSIS (Lewis et al. 2020) estimate that in 2020 cybercrime cost the world economy just shy of $1 trillion, which other studies (Morgan 2020) claim underestimates the true cost by a factor of six. HP (2021) reported a 100% increase in attacks perpetrated by nation-states from 2017 to 2020. Purplesec (2022) found that the number of malware infections had grown sixty-five times larger since 2009.[1] Other existing attack tallies show consistent yearly increases beginning in the 2000s, usually at an accelerating rate. The Federal Bureau of Investigation's (FBI) annual Internet Crime Complaint Center (IC3) reports (2022) show a general uptrend in both reported internet fraud and financial losses starting in 2001. While not every year at the IC3 showed an increase in cybercrime,[2] reported complaints and losses are seventeen and 388 times higher, respectively, today than in 2001. The "floor" of cyber conflict, in this sense is far above zero. In all, attacks ranging from amateur email scams to business-breaking ransomware and from simple typosquatting to staggeringly complex nation-state espionage campaigns continue to grow more common.

All these quotes and quantities illustrate one of cybersecurity's enduring puzzles: over the past decade, while cybersecurity seems more challenging and its failures more numerous and costly every year, the oft-prophesized cyber Armageddons, Pearl Harbors, and Skynets never seem to materialize. Chaos and conflict in cyberspace are persistent but not catastrophic; costly and harmful, yet strongly resistant to reduction; expanding in scale, but not in scope. In the public record, cyber insecurity has generally remained bounded between a *ceiling* and a *floor*, creating a narrow, yet expensive and quite unpleasant, band of insecurity. Other scholars (Fischerkeller et al. 2022) recognize this parameterized nature

DOI: 10.4324/9781003425304-9

of conflict in the domain and have integrated it into broader theories explaining within-domain behavior.

A variety of approaches in cybersecurity literature try to explain the phenomenon of cyber conflict as a product of the unique conditions of cyberspace – the pace of technological change, shaping effects of computing and networks, the man-made nature of the domain, and more. Instead of a technologically determined framing, the choice to protract, and the resulting stability of insecurity, is a product of natural economic and political incentives for organizations engaged in lower-intensity conflict, irrespective of the conditions of cyberspace. This chapter draws on the rich and relevant literature on irregular violence to illustrate and explain this mechanism. This category of conflict and malicious behavior is categorized primarily by violence below the threshold of war, including a spectrum of behavior from organized criminality to irregular warfare, conducted by entities organized primarily for the execution of this type of behavior. Studying this body of cases and related literature suggests an answer to the puzzle of the stability of insecurity in cyberspace: strategic protraction.

Various types of conflict, violence, and malicious activity outside of the cyber domain, from organized crime to irregular conflict, are characterized by an intentional protraction of that violence. In these models, violence is a means to create and sustain a period of insecurity rather than to achieve decisive strategic outcomes.

A recurring feature of this insecurity as ends (vs. decisive act) logic is it bedevils opposing strategists and policymakers. This logic reorients the discussion surrounding the degree of conflict from a focus on escalation, or the ceiling, to a focus on the position of bands themselves, the height of this box defined by ceiling and floor above true peace and security. This should refocus analysts on the incentives driving participation in and perpetuation of violence.

This chapter develops and explores these parallels through examples of irregular violence and malicious cyber activity which exploit existing opportunities to benefit created by the vulnerability and accessibility of the domain, like ransomware and kidnapping in areas of insecurity, as well as examples of created spaces for such exploitation by peer or near-peer competitors, like resource wars and cyber espionage. The chapter concludes with a handoff to future work and brief address of theoretical implications.

Defining the Bounds

The Ceiling

The *ceiling* is an upper bound on the intensity of observed cyber conflict, it is the line across which the likelihood of an action carrying great costs and small benefits grows quickly. This bound is most often located, in relation to cyber conflict, to where cyber effects generate direct kinetic effects. Precious few incidents have crossed that threshold – and even the most notorious examples there, NotPetya and Stuxnet, seem riddled with miscalculation or operational error. Even within the still ongoing Russo-Ukrainian war, cyber effects have proven more useful as shaping and logical, rather than decisive and kinetic, effects (Cattler and Black 2022; Schroeder et al. 2022; Schroeder and Dack 2023; Bateman 2022; Lewis 2022). While cyber conflict mostly remains below the ceiling, that level is not universal: it varies across domains and conflicts, and different actors make calculations based on their context. All actors can choose not to take action that they perceive as crossing this threshold if they determine it is likely to invite retaliation or escalation or if the desired effect can be delivered more efficiently. James Lewis's work (2018) on the political and strategic constraints of cyberattacks describes these forces well. At the most basic level, actors have legitimate fears about retaliation for escalatory cyber behavior and deliberately work to avoid conventional retaliation or even severe nonmilitary responses. Lewis also notes that many of the most extreme uses of cyber capabilities are simply "of limited benefit to [the perpetrators'] goals." (Lewis 2018). Cyber might simply be less useful for actors trying to achieve overt destruction, especially in times of open conflict outside the cyber domain – bombs are as good at disrupting power grids, manufacturing bases, and supply chains as cyberattacks, and they are generally cheaper, more reliable, and easier to contain.

The ceiling is an interplay between two or more actors, and their actions are only one part of the decision to determine whether it is viewed or responded to as "escalatory." The ceiling is only occasionally tested as actors seek room to maneuver and prod their adversaries. Researchers (Kushner 2021; AGCS Global 2016) have noted a reliable "return to the mean" (Fischkeller and Harknett 2019b, pp. 278 and 286) in those rare cases where an actor oversteps the bounds, too – when they misjudge cost and benefit systemically (as opposed to simply losing a potentially beneficial exchange), the consequences help them locate the ceiling rapidly (U.S. Cyber Command 2018).[3] Fischerkeller and Harknett (2019a) describe this testing as a process of tacit bargaining, by which adversaries implicitly agree upon ceiling, bounding the "competitive space short of armed conflict" (Fischkeller and Harknett 2019a) through repeated interaction.

Understandably, this upper threshold of escalation is much discussed and debated. And though important, the contextual nature of the cyber conflict ceiling means that such a focus is insufficient to understand the broader characteristic of insecurity across cyberspace. No actor's primary purpose in executing an operation, cyber or otherwise, is to avoid attention. Avoiding detection or reprisal is secondary. Actors primarily seek to change their environment and benefit in accordance with their strategic goals, that is, they pay far more attention to the floor.

The Floor

The *floor* constitutes the lower bound of conflict intensity, where malicious actors can reap the benefits of insecurity with the smallest possible input and risk. The floor is necessarily above zero, or perfect peace and stability, and the distance above that baseline can further increase as greater insecurity is normalized, or reduced, as it is problematized. In the cyber domain, some degree of exploitation is virtually guaranteed, given the clear opportunity to benefit, largely informed by the accessibility of offensive tools and the widespread vulnerability in cyberspace. In the context of cybercrime, for example, a Deloitte study (2018) found that penetration tools costing an average of just $3,800 a month could net cybercriminals $1 million over the same timeframe – such an impressive return that one should always expect some attempts to profit. According to FBI (2022) statistics, low-effort phishing scams were among the top five methods reported by victims of Internet crime from 2016 to 2020. However, the true figures are likely much higher, as estimates (Evans and Scott 2017) hold that less than a fourth of all fraud offenses are reported to law enforcement. These tactics caused financial losses of almost half a billion dollars in 2020 (FBI 2021), and only a very small percentage of scammers are arrested and prosecuted because of the challenges of identifying perpetrators, the sheer amount of criminal activity, and the noncooperation of some states with extradition (Eoyang 2018). Actors in the cyber domain require tools and some proficiency to execute attacks, but the barrier to entry for basic acts of aggression is extremely low, and access to these capabilities is projected to increase over time (Segal 2015).

Offensive capabilities are increasingly available for purchase, and online communities can provide necessary technical expertise at the click of a search button (Knapp and Boulton 2006). The growing commercialization of cyber criminality seems here to stay. Some authors of malware market their products internationally and offer customization and twenty-four-hour support services (Bartz 2010). Access-as-a-service companies sell complete offensive cybersecurity capabilities with user training and operational support (DeSombre et al. 2021). Freely available tools without malicious intent, such as Google Maps, social media sites, generative machine learning services, and the Shodan search engine, can also enable acts of aggression (Van Niekerk and Maharaj 2013; Burney et al. 2017). In cyberspace, there is no such thing as perfect defense – there is simply too much money to be made, too vulnerable an attack surface, too much access to the means to compromise it, and too little risk in trying.

The band of malicious activity bounded by this ceiling and floor describes an activity that is individually non-escalatory and that, to some degree, is affirmed by states as an acceptable degree of insecurity. The distance between the floor and absolute peace determines what kind of exploitative activities are possible and rewarding, while the ceiling acts as a bound through an inferred promise of reaction or reprisal. Within this space, actors have proven able to expand the scale of their malicious operations and reap significant benefits over time, creating powerful incentives to sustain that insecurity.

Protraction

Within irregular violence, entities engaging in such behavior operate on self-perpetuating incentives. Insecurity, within the bounds of the ceiling and floor, is the optimal environment for operation, so the stability of this insecurity is of strategic importance. Actors deliberately choose to protract their operations within this bound rather than seek a conventional, decisive victory because the benefit derived from that insecurity is an end to itself. In their study of insurgency and counterinsurgency in the 21st century, Dr. Stephen Metz and LTC Richard Millen (2004) describe protraction as the choice to "postpone decisive action" as a way to ensure continued operation. This protraction helps sustain a low but appreciable level of insecurity that enables rent-seeking, continued intelligence access, and other activities with strategic benefit and without the costs of a "new" conflict or increased attention. Protraction is more than just avoiding annihilation – fully realized as a strategy, it is creating and preserving an exploitable, insecure environment that allows an actor to benefit continuously throughout a conflict, even if just marginally. Insecurity breeds the opportunity for actors to benefit so long as that insecurity persists, and how they proceed is a strategic choice. In both irregular violence and malicious cyber activity, this protraction is visible in the operations of comparatively weak actors seeking to exploit opportunities and of peer or near-peer competitors who create space for opportunity.

Exploiting Chaos

Protraction as a strategy tends to benefit actors fighting at a disadvantage: smaller actors engage in low-level malicious activity, attempting to hide as an indistinguishable part of the general insecurity of an environment, and so preserve their ability to operate and benefit continuously. Hirschleifer's *Paradox of Power* (1991, pp. 177–200) tries to explain through economics the successes of smaller actors in a wide variety of conflicts, from military theatres to corporate takeovers. In a prolonged conflict, his framework describes the successes of smaller actors as a tax on larger actors, extracting as much benefit as possible without provoking an overwhelming response from the larger to disrupt, permanently, the smaller. Often, these benefits exceed what would be possible in a secure, conflict-free interaction.

Ransomware and kidnap-for-ransom operations illustrate the taxation model well. Shortland's (2019) extensive work on kidnap-for-ransom economics describes the taxation regime and vocabulary established by enterprise-scale kidnapping regimes like FARC. In the Colombian example, the benefits FARC derived from continued insecurity were principally financial, allowing the group to sustain its core political purpose. Developing a reputation for both kidnapping and returning hostages reliably "encourage[s] 'tax' compliance without driving away economic activities that fund the mafia or rebel organization," Shortland (2019, p. 15) writes.

There is a critical, insecure sweet spot in which systematic kidnapping is possible, but businesses will still try to operate, ransom insurers will still provide coverage, and hostages will be safely returned for payment. This is protraction in practice, and Shortland (2019, p. 147) goes on to describe how enterprises operating in Colombia treat paying the ransom as just another cost of doing business, no different from missed deadlines, maintenance repairs, or production delays. These conflict environments stem from the drawn-out insecurity between the bounds of a ceiling and floor: bad enough to exploit but not to incur an existentially threatening crackdown and risk an end to the fruitful insecurity.

Ransomware gangs are a model of this behavior, targeting entities with large capacity to pay but not so much visibility that it might draw an overwhelming response, striving to demand as much money as possible from their victims, but not so much that nonpayment is preferred. These gangs are essentially market participants, trying to ensure their own continued profitable operation. The dynamic directly mirrors kidnap-for-ransom markets, where insurers play a direct role in price-setting and where kidnappers strive to maintain an equilibrium of insecurity – great enough that they can flout the rule of law but not so great that brokers will refuse coverage, causing businesses to withdraw entirely from an area (Shortland 2019, p. 15). In both kidnapping and ransomware, malicious actors must preserve some semblance of reputability in order to benefit from the insecure environments they help create, perpetuate, and exploit. Jun's work (2021) on the logic of ransomware formalizes this dynamic in a game-theoretic model, where attackers need to demonstrate both malicious capability and good-faith efforts to restore data upon payment, leading perpetrators to provide surprisingly responsive "customer service" out of an explicit desire to maintain a reputation for reliability and professionalism. In the short term, for example, it could be more profitable to accept a paid ransom and then still use that company's data to reap further profit, but in the long-term such behavior would lessen the incentive of future victims to pay up at all.

The uptick in major ransomware attacks in 2021 illustrated the costs actors face when the intensity of the conflict they're waging rises too far above the floor – specifically, the ransomware attacks against both Kaseya software and JBS meat plant by the Russian-affiliated gang REvil (Menn 2021). The Kaseya operation affected up to 1,500 organizations, exploiting widely used IT-management software, and the attack on the meat supplier JBS forced the temporary closure of all of the company's facilities within the United States (Batista et al. 2021). The incidents highlighted the dangers of ransomware, particularly to consumers, who faced price hikes (Creswell 2021) and bouts of panic buying (Rapier 2021). The attacks also entered the news cycle on the heels of the Colonial pipeline shutdown (Bunge 2021). When asked how much ransomware money would be enough, one REvil operator in March 2021 said, "For me personally, there is no ceiling amount. I just love doing it and making a profit from it" (Smilyanets 2021). REvil went under the radar soon after, and much analysis implied their disappearance was an attempt to escape the spotlight (Tidy 2022). When the group returned to business a few months later, a coalition of states gained access to their systems, obtained a universal decryption key, and took the gang offline (Menn and Bing 2021; Arntz 2021). REvil, long able to operate in the shadows, had flown too close to the sun, incurring a state-backed response, including a cross-U.S. government operation and arrests of alleged REvil gang members by the Russian government, that the group was unable to endure (Menn and Bing 2021).

Creating Space for Competition

A whole class of literature on irregular conflict discusses resource conflicts – efforts to obtain direct material benefit from periods of insecurity in a physical space. This literature holds fruitful insights for the cybersecurity domain and the decision to protract. The widespread vulnerability, and accessibility, of digital systems can be understood in the terminology of resource conflict literature as an environment with a wealth of diffusely distributed resources. Resource abundance, according to research from Katharina Wick and Erwin H. Bulte (2006), frequently leads actors to choose "a re-allocation of effort from production toward rent-seeking or conflict" especially in an economy endowed with a diffuse resource.

This type of irregular violence reaps greater benefits from conflict than its resolution. In *Useful Enemies*, David Keen (2012, p. 8) finds echoes of this logic in modern conflicts across Africa, arguing that "a great many wars are resistant to ending for the simple (but hidden) reason that powerful actors (both local and international) do not want them to end." Some actors with the capability to benefit from escalating above the irregular conflict ceiling choose not to. Following an attempted coup in 1991, Sierra Leone descended into a messy, decade-long civil war among many competing groups in a tangled web of strategy, ideology, and resource extraction (Keen 2012, pp. 18–22). Even actors with superior resources or military capability avoided direct conflict with their opponents as they contested the wealth provided by diamonds and other resources (Keen 1998). The insecurity in the country made the exploitation of those resources all the easier (Kaldor 2017). Throughout the war, various actors worked to preserve and exploit this insecurity, even those at an asymmetric advantage. A 2000 UN Panel (2000, para 248) noted an overwhelming number of reports of soldiers providing weapons to rebel soldiers in exchange for cash, diamonds, and other goods (International Crisis Group 2001). The conflict's parties felt the benefits of participation outweighed those of resolution, so they extended the conflict.

In cyberspace, the strategies of U.S. adversaries make clear the benefits of protracted, low-grade conflict. The industrial cyber espionage policies of the Chinese government, according to a former U.S. Department of Justice Attorney General's statement to the U.S. Senate hearing on Chinese espionage, reflect a long-term strategy of strategic accumulation through espionage. Through cyber operations, according to the former Attorney General's statement, the Chinese government has "gained unauthorized access to a wide range of commercially valuable business information, including trade secrets, technical data, negotiating positions, and sensitive and proprietary internal communications" (Anon 2018) driving increased domestic technological development while burdening and restricting the U.S. market.

Estimates (Anon 2017) put the cost of intellectual property theft from U.S. companies between $180 and $540 billion per year. In 2007, hackers breached the networks of military subcontractor Lockheed Martin (Nasaw 2009), compromising data from the U.S. Joint Strike Fighter program and the still-in-development F-35. Subsequent leaks from NSA contractor Edward Snowden, suggested the Chinese government was responsible, using the data to inform their new J-31 stealth fighter and J-20 fighter jet (Gady 2015; Gorman et al. 2009). Though the pace of Chinese cyber intrusions decreased following U.S.–China cyber agreements in 2015 and 2016 (FireEye 2015), that rate seems to have returned to the mean (Gazis 2018).[4] China's cyber espionage and intellectually-property-theft operations help its government chip away at the economic, technological, and military gaps between itself and the United States. Notably, the short period of "cooperation" after the U.S.–China cyber agreements allowed the PRC to reset the cadence and intensity of its operations and preserve its ability to continue espionage operations – protraction in practice. Over time, these incursions achieve significant strategic effects, much like insurgencies. Former NSA Director and U.S. CYBERCOM General Keith Alexander famously described industrial cyber espionage and intellectual property theft as "the greatest transfer of wealth in history" (Rogin 2019) precisely because of their accumulated effects, rather than the result of any single incident.

In cyberspace, just as in these varying forms of irregular violence, low and slow can outperform loud and fast over a long enough time, regardless of actor size. A strategy of protraction – creating and preserving an insecure, exploitable environment – is key to leveraging that.

Moving Forward

In any ecosystem, and perhaps especially one characterized by the widespread vulnerability found in the modern digital ecosystem, some degree of insecurity and exploitation must be expected. This band of activity is not fixed but is the result of responsive incentives, the interplay of participant actions and expectations which contribute to a stable insecurity. In this insecurity, malicious activity prospers without escalating to a dangerous level which might draw an overwhelming response and resulting loss of benefit.

The United States and its allies, in their pursuit of a more secure cyber domain, must address not only the upper "limit" of cyber malicious activity, but seek to push the entire band downward, reducing the range of acceptable insecurity. Policies meant to engage and tacitly bargain with the largest or most threatening adversaries, working to reduce the maximum potential consequences of offensive cyber activity, must not ignore the position of the band itself and refocus on the floor as an equal measure of conflict intensity and outcomes. Single cyber policies do not have consistent effects against all behaviors from the ceiling to the floor of conflict. The United States and close allies must work to directly address the incentives to protract conflict at low intensities.

Crucially, this reshaping of the insecurity of cyberspace cannot be considered as an intent to deter or eradicate all malicious activity, nor to squeeze operators acting near this lower bound "upward" to chase moving incentives. The United States and allied cyber defense and cyber-oriented law enforcement must shift their thinking to match the protracted, long-term viewpoint of malicious actors operating in this domain. This is as challenging in cyberspace as in the varying forms of irregular violence these political entities have been engaged in over the past 100 years, and it asks much of analysts to resist mirror-imaging their own strategic logic of conflict. As adversaries seek to exploit or create opportunities in this space, the United States must make these opportunities more and their benefits far harder to obtain. This is both a function of reducing the widespread vulnerability in digital technologies that enable such low-cost insecurity and explicit efforts to counter strategies of protraction.

More work remains to be done, and this piece has barely plumbed the great depths of the rich literature covering different forms of irregular violence. Future work should seek to specify the individual decision-making logics of small states with regard to their activities in cyberspace, understand the variations in strategy between different kinds of criminal groups, and address new entrants to the conflict in the form of traditional non-state armed groups adopting cyber capabilities. The sustained insecurity of the cyber domain is a result of both structural factors and deliberate choices. Understanding more deeply the system of incentives that drive a wide variety of actors to engage in this conflict and protract cyber insecurity is essential for the United States and its allies to reshape the systemic incentives buttressing the stability of insecurity in cyberspace.

Notes

1 Worryingly, these estimates come from vendors of products and services meant to stem the cyber losses – in the United States, there is no central source to provide cyber statistics or analysis for the public.
2 2006 and 2007 were flat compared to 2005 before taking off again.
3 Journalists and politicians have described some attacks as crossing a certain threshold of conflict – in 2010 the Stuxnet worm, the first to inflict physical damage, was discovered in the SCADA systems of an Iranian centrifuge plant, and in 2015 the first cyber-induced electrical blackouts afflicted Ukraine – but incidents like these have not yet spiraled into sustained escalation.
4 FireEye and the Department of Justice assess the decline between 2013 and 2016 may have been attributed to the Chinese refining their tactics and techniques. This rebound may simply reflect the United States' increased ability to detect and identify these intrusions. See also Segal (2016).

References

AGCS Global, 2016. *Cyber Attacks on Critical Infrastructure*. Available from: www.agcs. allianz.com/news-and-insights/expert-risk-articles/cyber-attacks-on-critical-infrastruct ure.html

Anon, 2017. *Update to the IP Commission Report*. Washington, DC: The National Bureau of Asian Research.

Anon, 2018. *Findings of the Investigation into China's Acts, Policies, and Practices Related to Technology Transfer, Intellectual Property, and Innovation under Section 301 of the Trade Act of 1974*. Washington, DC: Office of the United States Trade Representative, Executive Office of the President.

Arntz, P., 2021. [Updated]Revil Ransomware Disappears after Tor Services Hijacked. *Malwarebytes*, 19 October. Available from: www.malwarebytes.com/blog/news/2021/10/ revil-ransomware-gang-disappears-after-tor-services-hijacked

Bartz, D., 2010. Analysis: Top Hacker "Retires"; Experts Brace for His Return. *Reuters*, 29 October. Available from: www.reuters.com/article/us-hackers-zeus-idUSTRE69S54Q2 0101029

Bateman, J., 2022. Russia's Wartime Cyber Operations in Ukraine: Military Impacts, Influences, and Implications. *Carnegie Endowment*, 16 December. Available from: https://carnegieendowment.org/2022/12/16/russia-s-wartime-cyber-operations-in-ukraine-military-impacts-influences-and-implications-pub-88657

Batista, F., Hirtzer, M., and Dorning, M., 2021. JBS Cyber Hack: Meat Supplier Shuts Down Some Slaughterhouses after Attack. *Bloomberg*, 31 May. Available from: www.bloomberg.com/news/articles/2021-05-31/meat-is-latest-cyber-victim-as-hackers-hit-top-supplier-jbs#xj4y7vzkg

Bunge, J., 2021. JBS Paid $11 Million to Resolve Ransomware Attack. *The Wall Street Journal*, 9 June. Available from: www.wsj.com/articles/jbs-paid-11-million-to-resolve-ransomware-attack-11623280781

Burney, A., Asif, M., Abbas, Z. and Burney, S., 2017. Google Maps Security Concerns. *Journal of Computer and Communications*, 6(1), 275–283.

Cattler, D. and Black, D., 2022. The Myth of the Missing Cyberwar. *Foreign Affairs*, 6 April. Available from: www.foreignaffairs.com/articles/ukraine/2022-04-06/myth-missing-cyberwar

Creswell, J., Perlroth, N. and Scheiber, N., 2021. Ransomware Disrupts Meat Plants in Latest Attack on Critical U.S. Business. *The New York Times*, 1 June. Available from: www.nytimes.com/2021/06/01/business/meat-plant-cyberattack-jbs.html

Dal Cin, P., Fox, J., Sidhu, H. and Nunn-Price, J., 2021. *State of Cybersecurity Report 2021: 4th Annual Report*. Accenture. Available from: www.accenture.com/us-en/insights/security/state-cybersecurity

Deloitte, 2018. Deloitte Puts the Spotlight on the Cost of Cyber-Crime Operations in New Threat Study. *Deloitte*, 14 December. Available from: www2.deloitte.com/us/en/pages/about-deloitte/articles/press-releases/deloitte-announces-new-cyber-threat-study-on-criminal-operational-cost.html

DeSombre, W., Shires, J., Work, JD., Morgus, R., Howell O'Neal, P., Allodi, L., and Herr, T., 2021. Countering Cyber Proliferation: Zeroing in on Access-as-a-Service. *Atlantic Council Cyber Statecraft Initiative*, 1 March. Available from: www.atlanticcouncil.org/in-depth-research-reports/report/countering-cyber-proliferation-zeroing-in-on-access-as-a-service/

Eoyang, M., 2018. *To Catch a Hacker: Toward a Comprehensive Strategy to Identify, Pursue, and Punish Malicious Cyber Actors,* s.l.: Third Way.

Evans, M. and Scott, P., 2017. Fraud and Cyber Crime Are Now the Country's Most Common Offences. *The Telegraph*, 19 January. Available from: www.telegraph.co.uk/news/2017/01/19/fraud-cyber-crime-now-countrys-common-offences/

FBI, 2021. *Internet Crime Report 2020*. Available from: www.ic3.gov/Media/PDF/AnnualReport/2020_IC3Report.pdf

FBI, 2022. *Internet Crime Complaint Center(IC3): Annual Reports*. Available from: www.ic3.gov/Home/AnnualReports

FireEye, 2015. APT30 and The Mechanics of a Long-running Cyber Espionage Operation. *Mandiant*. Available from: www.mandiant.com/resources/reports/red-line-drawn-china-recalculates-its-use-cyber-espionage

Fischerkeller, M. P., Goldman, E. O. and Harknett, R. J., 2022. *Cyber Persistence Theory: Redefining National Security in Cyberspace*. 1st ed. New York, NY: Oxford University Press.

Fischkeller, M. P. and Harknett, R. J., 2019a. What Is Agreed Competition in Cyberspace? *Lawfare*, 19 February. Available from: www.lawfaremedia.org/article/what-agreed-competition-cyberspace

Fischkeller, M. P. and Harknett, R. J., 2019b. Persistent Engagement, Agreed Competition, and Cyberspace Interaction Dynamics and Escalation. *The Cyber Defense Review*, Special Edition 2019, 267–287. Available from: https://cyberdefensereview.army.mil/Portals/6/CDR-SE_S5-P3-Fischerkeller.pdf

Gady, F.-S., 2015. New Snowden Documents Reveal Chinese behind F-35 Hack. *The Diplomat*, 27 January. Available from: https://thediplomat.com/2015/01/new-snowden-documents-reveal-chinese-behind-f-35-hack/

Gazis, O., 2018. U.S.' Top Spy-Catcher: China Brings 'Ungodly Resources' to Espionage. *CBS News*, 9 September. Available from: www.cbsnews.com/news/ncsc-director-says-china-is-the-largest-threat-to-national-security/

Gorman, S., Cole, A. and Dreazan, Y., 2009. Computer Spies Breach Fighter-Jet Project. *The Wall Street Journal*, 21 April. Available from: www.wsj.com/articles/SB124027491029837401

Hirshleifer, J., 1991. The Paradox of Power. *Economics and Politics*, 3(3), 177–200. https://doi.org/10.1111/j.1468-0343.1991.tb00046.x

HP Wolf Security, 2021. Nation States, Cyberconflict and the Web of Profit: HP Threat Research. *HP Threat Research Blog*, 8 April. Available from: https://threatresearch.ext.hp.com/web-of-profit-nation-state-report/

International Crisis Group, 2001. Sierra Leone: Time for a New Military and Political Strategy. *International Crisis Group*, 11 April. Available from: www.crisisgroup.org/africa/west-africa/sierra-leone/sierra-leone-time-new-military-and-political-strategy

Jun, J., 2021. The Political Economy of Ransomware. *War on the Rocks*, 2 June. Available from: https://warontherocks.com/2021/06/the-political-economy-of-ransomware/

Kaldor, M., 2017. *New and Old Wars*. 3rd ed. Cambridge, UK: Polity Press.

Keen, D., 1998. The Economic Functions of Violence in Civil Wars. *The Adelphi Papers*, 38(320), 7–88.

Keen, D., 2012. *Useful Enemies: When Waging Wars is More Important Than Winning Them*. 1st ed. New Haven: Yale University Press.

Knapp, K. J. and Boulton, W. R., 2006. Cyber-Warfare Threatens Corporations: Expansion into Commercial Environments. *Information Systems Management*, 23(2), 76–87. https://doi.org/10.1201/1078.10580530/45925.23.2.20060301/92675.8

Kushner, D., 2021. The Real Story of Stuxnet. *IEEE Spectrum*, 26 February. Available from: https://spectrum.ieee.org/the-real-story-of-stuxnet

Lewis, J. A., 2018. *Rethinking Cybersecurity Strategy, Effect, and States*. Center for Strategic & International Studies. Available from: https://csis-website-prod.s3.amazonaws.com/s3fs-public/publication/180108_Lewis_ReconsideringCybersecurity_Web.pdf

Lewis, J. A., 2022. *Cyber War and Ukraine*. Center for Strategic & International Studies. Available from: https://csis-website-prod.s3.amazonaws.com/s3fs-public/publication/220616_Lewis_Cyber_War.pdf

Lewis, J. A., Malekos Smith, Z. L. and Lostri, E., 2020. *The Hidden Costs of Cybercrime*. Center for Strategic and International Studies, McAfee Security.

Menn, C. and Bing, J., 2021. Exclusive Governments Turn Tables on Ransomware Gang Revil by Pushing It Offline. *Reuters*, 21 October. Available from: www.reuters.com/technology/exclusive-governments-turn-tables-ransomware-gang-revil-by-pushing-it-offline-2021-10-21

Menn, J., 2021. Kaseya Ransomware Attack Sets off Race to Hack Service Providers – Researchers. *Reuters*.

Metz, S. and Millen, R., 2004. *Insurgency and Counterinsurgency in the 21st Century*. Carlisle, PA: US Army War College Press. www.jstor.org/stable/resrep11453

Morgan, S., 2020. Global Cybercrime Damages Predicted to Reach $6 Trillion Annually by 2021. *Cybercrime Magazine*, 9 November. Available from: https://cybersecurityventures. com/annual-cybercrime-report-2017/

Nasaw, D., 2009. Hackers Breach Defences of Joint Strike Fighter Jet Programme. *The Guardian*, 21 April. Available from: www.theguardian.com/world/2009/apr/21/hackers-us-fighter-jet-strike

Panel of Experts Appointed Pursuant to UN Security Council Resolution 1306, 2000. *Report of the Panel of Experts in relation to Sierra Leone*, paragraph 248: Report of the Panel of Experts.

Purple Security, 2022. 2022 Cyber Security Statistics Trends & Data. *Purplesec*. Available from: https://purplesec.us/resources/cyber-security-statistics/

Rapier, R., 2021. Panic Buying Is Causing Fuel Shortages along the Colonial Pipeline Route. *Forbes*, 11 May. Available from: www.forbes.com/sites/rrapier/2021/05/11/panic-buying-is-causing-gas-shortages-along-the-colonial-pipeline-route/

Rogin, J., 2019. NSA Chief: Cybercrime Constitutes the 'Greatest Transfer of Wealth in History. *Foreign Policy*, 9 July. Available from: https://foreignpolicy.com/2012/07/09/nsa-chief-cybercrime-constitutes-the-greatest-transfer-of-wealth-in-history

Schroeder, E., and Dack, S., 2023. *A Parallel Terrain: Public-private Defense of the Ukrainian Information Environment*. Washington, DC: The Atlantic Council. Available from: www.atlanticcouncil.org/in-depth-research-reports/report/a-parallel-terrain-public-private-defense-of-the-ukrainian-information-environment/

Schroeder, E., Scott, S. and Herr, T., 2022. Victory Reimagined: Toward a More Cohesive US Cyber Strategy. *Atlantic Council Cyber Statecraft Initiative*, 14 June. Available from: www.atlanticcouncil.org/in-depth-research-reports/issue-brief/victory-reimagined/

Segal, A., 2015. The Code Not Taken: China, the United States, and the Future of Cyber Espionage. *Bulletin of the Atomic Scientists*, 69(5), 38–45. https://doi.org/10.1177/00963 40213501344

Segal, A., 2019. The U.S.-China Cyber Espionage Deal One Year Later. *Council on Foreign Relation*, 28 September. Available from: www.cfr.org/blog/us-china-cyber-espion age-deal-one-year-later

Shortland, A., 2019. *Kidnap: Inside the Ransom Business*. Oxford, UK: Oxford University Press.

Smilyanets, D., 2021. I Scrounged through the Trash Heaps... Now I'm a Millionaire:' an Interview with Revil's Unknown. *The Record (Recorded Future)*, 16 March. Available from: https://therecord.media/i-scrounged-through-the-trash-heaps-now-im-a-milliona ire-an-interview-with-revils-unknown

Sonatype, 2021. *2021 State of the Software Supply Chain; The 7th Annual Report on Global Open Source Software Development*. Available at: www.sonatype.com/hubfs/SSSC-Rep ort-2021_0913_PM_2.pdf?hsLang=en-us

Tidy, J., 2022. REvil Ransomware Gang Arrested in Russia. *BBC News*, 14 January. Available from: www.bbc.com/news/technology-59998925

U.S. Cyber Command, 2018. *Achieve and Maintain Cyberspace Superiority: Command Vision for U.S. Cyber Command*. Available from: https://assets.documentcloud.org/documents/4419681/Command-Vision-for-USCYBERCOM-23-Mar-18.pdf

Van Niekerk, B. and Maharaj, M., 2013. Social Media and Information Conflict. *International Journal of Communication*, 7, 1162–1184.

Wick, K., and Bulte, E. H., 2006. Contesting Resources: Rent Seeking, Conflict and the Natural Resource Curse. *Public Choice*, 128(3/4), 457–476. www.jstor.org/stable/25487568

9   Digital IEDs on the Information Highway

PSYOPS, CYBER, and the Info Fight

Chaveso L. Cook

Introduction

As the art of war changes, so does its science. The 21st century's revolution in military affairs saw countries like the United States rise to have limited near-peer competitors in conventional military power. To challenge the king of the hill, many adversaries responded by progressively turning to irregular and asymmetric methods to engage in competition and conflict (Lin and Kerr 2021). The 1990s internet explosion eventually caused an eruption of connectivity previously unseen, birthing the cyber arena of competition and conflict (Cook 2022). Cyber is a domain that offers many power seekers and competitors an opportunity to counter conventional military advantages, including technological overmatch, especially when it comes to the realm of information warfare. In turn, the cyber domain has become a key means to conduct information warfare, often through the distribution of propaganda and extremist ideologies. As such, radical ideology and its associated acts of both domestic and foreign terrorism are a serious threat to the global community (Cook and Collins 2021).

Ideology, radical or otherwise, is a manifestation of deeper beliefs based upon intensely held but rarely understood underlying assumptions. A bullet may kill an extremist, but it will not kill their ideology; that is, "bullets do not kill ideas; a "hot" war against an idea is destined to be a losing prospect" (Staton 2020). To wit, words, pictures, tweets, texts, posts, videos, links, websites, and the like have all become much cheaper than bullets (Duggan 2014). Perhaps a 9/11-type event is not as likely to happen again in the form of mass transit – planes, trains, or other vehicle-born weaponry. If that be the case, what one could surely argue is that it is much more likely to happen through mass media – cultural exploitation, political chess, electoral manipulation, cyber intrusion, and social mal-influence. Most dangerously, the event will likely be just beneath the surface, more IED than WMD.

The ubiquity of the internet and social networking involves exponential growth of a globally connected culture. Life now has a digital touch, where we log in to a thriving online society that parallels and mirrors the community. Cyber-based influence has undoubtedly crept into our daily lives, shaping many parts of our experience to include war. Cilluffo and Clark (2014) adroitly state that "the ability to use cyberspace to create advantages and influence events in all other operational

DOI: 10.4324/9781003425304-10

environments and across instruments of power will ensure significant advantage" (p. 112). The joint community has acknowledged this fact, making information the newest joint function (Paul 2020). Information warfare is now decidedly a part of modern conflict (Ventre 2016). Consequently, a comprehensive understanding of the web is critical for the defense of any nation, as the internet has rendered our borders borderless.

As a manifestation of Moore's (1965) Law, technology has advanced exponentially, and the associated technological platforms have evolved even more (Bondyopadhyay 1998). These platforms need to be understood, as recognized in creating organizations like the U.S. Cyber Command (USCYBERCOM) in 2009. Today, USCYBERCOM's mission is to direct, synchronize, and coordinate cyberspace planning and operations to defend and advance national interests in collaboration with domestic and international partners and has become a full and independent unified combatant command (Cyber 2021).

However, if digital bullets and cyber IEDs are the weapons of choice for tomorrow's battlefield, then it may be the case that large, sprawling establishments like USCYBERCOM will not be best positioned to fight without parallel efforts that align with specially trained elements that focus on the human terrain. Winning in multi-domain operations of space, cyberspace, air, land, and sea will not occur without a critical understanding of the behavior and political wills of the enemy, particularly in the case of China and Russia (Czege 2020). The next age of war, cold or hot, will be waged on the terrain of hearts and minds. The goal of this chapter is to examine what smaller, more agile psychological operations (PSYOP) forces bring to the fight as distinct advantages regarding targeted online influence efforts, as well as illustrate their connection to the efforts of the cyber community.

Critical Context

The first recorded cyber-attack was not in 2008 or even 1998 – it was in 1988 (Gordon and Rosenbach 2021). A computer science graduate student at Cornell University hacked into the computers of the Massachusetts Institute of Technology and released a worm that ultimately affected Lawrence Livermore National Laboratory and NASA (FBI 2018). Emails were delayed for days. Important university functions, and eventually vital military processes, ground to a near halt – and this all occurred a year before the invention of the World Wide Web (FBI 2018). However, cybersecurity only remained a concern for computer geeks, hackers, and intelligence operatives until two decades later when another worm, known as Stuxnet, shut down centrifuges in Iran (Gordon and Rosenbach 2021). At the time, USCYBERCOM had just been established and was initially placed under the command and control of U.S. Strategic Command, an entity that still oversees the U.S. nuclear arsenal and some space operations (Cyber 2021). This command-and-control authority suggested that cyber operations were viewed as military actions in outer space or "analogous to nuclear conflict," and Secretary of Defense Robert Gates even went so far as determining that the new command "would not carry out so-called information operations designed to influence the perceptions, thoughts, or beliefs of foreign actors to serve U.S. strategy" (Gordon and Rosenbach 2021, p. 13).

Today, both state and non-state actors are heavily investing in the "informational sphere, placing their actions of communication, influence, propaganda, [and] psychological operations at the heart of their strategies" (Ventre 2016, p. xiii). The United States has evolved not only its DoD structure to address the challenges but is also reforming and adapting policy to keep pace (Gordon and Rosenbach 2021). Unlike conventional or nuclear weapons, governments do not control the internet – it is a network of networks, most of which is privately owned (Nye 2021). From individual hackers to nation states, cyber warfare activities can do everything from crippling economies to inciting political unrest (Atrews 2020). Schneider (2021) states that "cyberthreats erode the foundations of which markets, societies, governments, and international systems are built" (p. 26). Senior political and military leaders across the globe have repeatedly expounded on the importance of the information environment for military operations and declared it a priority (Paul 2020). Arguably, success against propaganda in the cyber domain hinges less on deftly maneuvering within the hypertext transfer protocol and more in the psychological battlespace, i.e., the "gray matter," or decision-making apparatus, of both the adversary and their populations. Therefore, we must see the internet as the means, not the ends. Perspective with precedent matters here. If strategists still believed that Carl von Clausewitz's (1984) idea that war is an extension of politics, while also accepting Naim's (2014) claims in his book "The End of Power" that power no longer resides exclusively (if at all) in states, institutions, or large corporations, then centers of gravity will remain located in the networks that structure society.

The information revolution has created new economic entities, ones predicated on streams of data and social networks and possessing at least as much power as other forms of organization. Dmitri Alperovitch (2021) argues that cyberspace "is not an isolated realm of its own, but an extension of the larger geopolitical battlefield" (p. 46). However, these efforts remain human endeavors animated by psychological functioning. As such, the fight of today and tomorrow is one of understanding minds, beliefs, and behaviors (Cowan and Cook 2018). To this end, Dave Stephenson observes, "exquisite understanding is more important than exquisite technology" (Amble 2020).

Before the advent of both USCYBERCOM and the U.S. Special Forces, the PSYOP practitioner (PSYOPer) shouldered the ability to understand, operate within, and influence populations (Cook 2014). In the informal PSYOP creed, Edward Rouse (2021) states that a PSYOPer has historically taken on many roles; inspirer, motivator, intimidator, and deceiver, to name a few. PSYOP soldiers today conduct Military Information Support Operations and are experts in mass media communication, using their unique skills to persuade, change, and influence foreign audiences. Psychological Operations have been officially defined in Defense Department (Department of the United States Army 2013) doctrine as "planned political, economic, military, and ideological activities directed towards foreign countries, organizations, and individuals in order to create emotions, attitudes, understanding, beliefs, and behavior favorable to the achievement of United States political and military objectives." In short, these operations are the functions of the DoD devoted to changing attitudes and behavior in foreign target audiences and are frequently described as propaganda outside the military (Cowan and Cook 2018).

Nations around the globe will continue to encounter foes who seek to conduct non-standard, unconventional, irregular type warfare. However, regardless of the methods that may be used by these adversaries, the ultimate objective is to change perception and opinion. After decommissioning the U.S. Information Agency, influencers have had no choice but to leverage the internet as a critical piece of infrastructure. The internet, especially social media, has become an integral part of the kill chain (Shallcross 2017). Nevertheless, to use this infrastructure effectively, an online information warfare practitioner must also be well-versed and well-practiced in changing the behaviors of the internet's human users (Cukier 2005).

Doctrinal Foundations and Challenges

The multiplicative efforts by organizations like USCYBERCOM and PSYOP elements create two challenges. From a DoD perspective, the first is tied to doctrine. At the inception of USCYBERCOM's *Joint Publication (JP) 3-13* (2014b) had an Information Operations Roadmap consisting of interrelated pillars: computer network operations (i.e., computer network attack, computer network defense, and computer network exploitation); PSYOP; electronic warfare; operations security; and military deception. Cyberspace is defined as "the global domain within the information environment consisting of the interdependent network of information technology infrastructures and resident data, including the internet, telecommunications networks, computer systems, and embedded processors and controllers" (DoD JP3-13 2014b). However, this definition lacks the cognitive, human element that the internet represents; this omission has adversely affected how the military organizes, trains, and utilizes its forces (DoD JP1-02 2014a).

The second challenge involves the velocity and volume of disinformation, propaganda, and threats to cyber security in the contemporary information space. The erosion of global borders is perhaps inversely proportional to the growth in internet usage. Contemporary life, therefore, has a ubiquitous digital component; increasingly, people around the globe log on to a thriving online society that mirrors their physical communities. Therefore, cyberspace and its influence have undoubtedly shaped all interactions, up to and including warfare, and technology has increased options for the antagonist as much as it has for protagonist (Rid and Hecker 2009). Those that "seize the key terrain of social media exploitation will have a strategic military advantage" (Duggan 2014, p. 68).

In turn, cyber-based influence has become a continually iterative and time-sensitive process. It is not only a question about who can message *first* (be "quick on the draw"), but also who can message *most often* (keep a sustained rate of fire). Shirky (2011) noted that "as the communication landscape gets denser, more complex, and more participatory, the networked population is gaining greater access to information, more opportunities to engage in public [sentiment] and an enhanced ability to undertake collective action" (p. 29). As much as this situation has been positive for global growth, in the modern operating environment "the nearly limit-less potential for strategic communication on the internet has [also] not gone unnoticed by terrorist organizations" (Gendon et al. 2009, p. 9).

More and more, there has been a belief that if "used preemptively, [online activities] could keep a conflict from evolving in a more lethal direction" (Gjelten 2013, p. 34). As electronic media dominate today's battlefield, the reality is that *all* future conflicts will contain cyber elements at all levels of warfare; regard-less of asymmetries in capabilities, usage of cyber components is "now a part of the strategic environment writ large" (Cilluffo and Clark 2014, p. 112). However, there are natural voids that PSYOP should fill. Leveraging social networks and the human terrain for targeted influence should be a paired responsibility of entities with similar efforts and goals, like USCYBERCOM and the PSYOP community, but with PSYOP in the lead.

Social Network Analysis (SNA) and Targeted Cyber Effects

Cyber-based threats directly jeopardize society's human networks – the bonds and links that people have as individuals, neighbors, and citizens" (Schneider 2021, p. 30). Practitioners – PSYOP and otherwise – of information operations have long studied and leveraged these social networks. At its core, a social network – whether face-to-face or web-based – is a map of relevant ties among participants in the net-work through nodes and links (Gendon et al. 2009, p. 8). Analysis of the informa-tion passing through the network empowers the strategic influencer with thematic guidelines to craft products that ensure messages appear indigenous in nature. SNA identifies primary information sources within the social network, thereby allowing friendly forces to target "influence brokers" and key communicators. Studying this information holistically can provide influence experts with the behavioral data required to execute effective, focused, timely, and decisive influence operations. In addition to baseline data, SNA can also be another tool with which to measure operational success or battlefield effectiveness.

Often, it is only through accessing primary network nodes (or key communicators) and, exploiting them as secondary dissemination platforms, that influence operations can effectively alter the behavior of a given target audience (TA) that receives most, if not all information, from the primary node. Understanding the utility of influence brokers within a social network relies relatively little on how a com-puter network's hardware is wired or how its software is implemented. The greatest should be placed on attaining an intricate understanding of the node's networks and what information they craft and pass to their networked consumers. Therefore, whereas cyber elements may provide access to a node or a TA, PSYOP elements working in concert should be the ones conducting actions that influence the net-work. It is through iterative analyses of these specific actions and their respective influence outcomes that provide what the influence expert needs to leverage for targeted influence.

Targeting is the act of selecting and prioritizing targets via operational requirements and capabilities and matching appropriate responses to them (Bourne

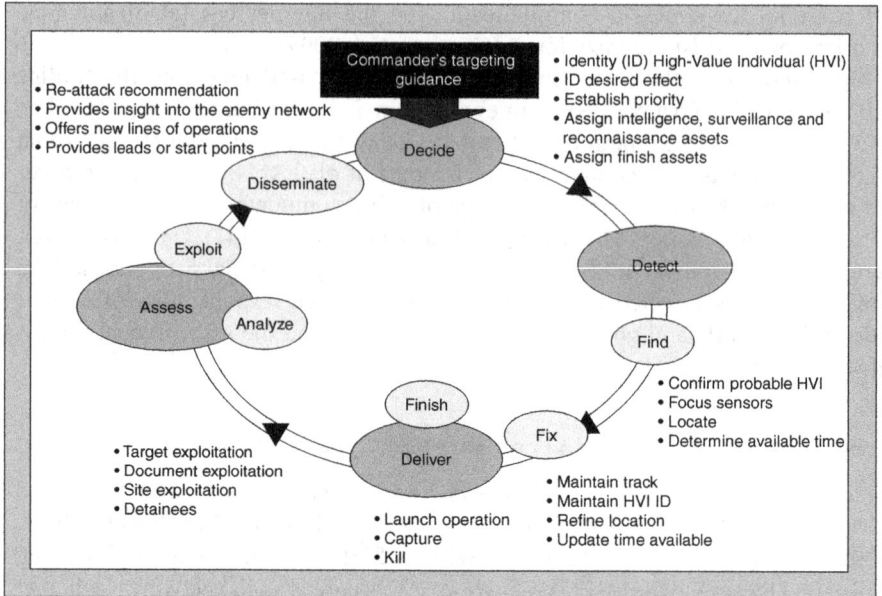

Figure 9.1 D3A and F3EAD (DoD FM 3-60 2015).

Source: Author's creation.

2019). Whereas the targeting cycle of F3EAD (*find, fix, finish, exploit, analyze, disseminate*) is consistent with the *decide, detect, deliver,* and *assess* (D3A) methodology (see Figure 9.1), F3EAD has grown in prominence as it provides maneuver commanders an additional tool to address certain targeting challenges, particularly those found in the cyber domain (DoD FM 3-60 2015b). As displayed in Figure 9.2, using the approach of the U.S. Army's seven-step PSYOP planning process overlaid onto a targeting cycle can help influence experts to advise information warfare efforts (Cook 2022).

During the seven-step process of *Planning (PLAN), Target Audience Analysis (TAA), Series Development (SDEV), Product Development and Design (PDEV), Approval (APP), Production, Distribution and Dissemination (PDandD),* and *Evaluation (EVAL)* PSYOPers are essentially in a targeting cycle (see Figure 9.2) (DoD FM 3-05.301 2015a). Starting with the intent and the desired end state, as given by the commander, to establishing baselines and initial assessments of measures of performance/effect, and moving through the rest of the process, the use of SNA inherently augments the targeting process (Brown 2012). Using this adapted targeting cycle, the inclusion of SNA into offensive and defensive operations can appropriately shift and direct how influence experts fire digital bullets and detect virtual IEDs, but it also can help translate the effects decision makers can achieve.

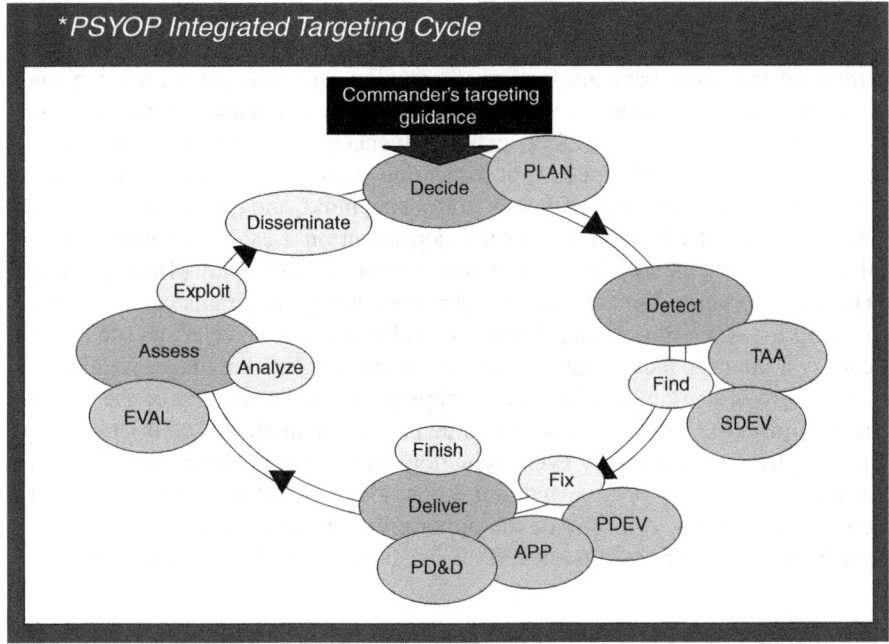

Figure 9.2 PSYOPS Process Overlaid on the Targeting Cycle.

Source: Author's creation.

The most profound SNA advantage is that it gives leaders the ability to glean real-time atmospherics and sentiment of TAs for targeting efforts – a critical component of the execution of successful and decisive operations with or without the traditional "boots on the ground" presence. SNA sheds "unprecedented light onto what people think and, more importantly, why they think it, as well as unparalleled access to those who see value in understanding [a population's] perspectives" (Bostick 2011, p. 17). Lin (2020) states that USCYBERCOM is on the information *delivery* side of any psychological effects. Through extensive experience in executing information operations, USCYBERCOM experts should be in sync with PSYOPers who have developed deep and broad psychological, anthropological, and cultural expertise on the information *content* side of creating psychological effects (Lin 2020). These operations run parallel and supplemental to existing processes within cyberspace and are not meant to replace existing functional capabilities.

Furthermore, cyber skills are seen as "hard skills," such as understanding, enumerating, and infiltrating network structure or managing malware and architectural vulnerabilities, whereas PSYOP skills are "soft skills," such as human psychology, cultural-linguistics, deception/ruse, influence tactics, marketing techniques, etc. (Lin 2020). SNA can certainly be enhanced by technology, but undoubtedly, understanding humans and our networks began long before the introduction of the internet to the battlespace. We fought under the duress of Axis Sally, Tokyo Rose, and Hanoi Hannah well before the cyber domain existed, yet cyberattacks today still rely on the same baseline tactics of "preying on weak points, sowing distrust, creating confusion and anxiety, and exacerbating hatred and misinformation" (Schneider 2021, p. 24). Therefore, cyber influencers must remember Widener's (2000) argument that "influence is not something you do to other people but rather that it is something that starts with how you shape, mold, and present information" (p. 5). Consequently, influence is greatly dependent on a calculated combination of plans, words, and actions before technology even comes to play (Turnley 2011).

There are certainly limits to computer programming; programs can be flawed from inception when interpreting human behaviors and may give errors (i.e., "sentiment" is a concept that is hard to capture in the English language, let alone a foreign language or slang). SNA, and an appropriate understanding of the human element, allows the internet influencer to be present in an area of conflict without the use of physical or possibly even technological actions, unlike a cyber-element who ostensibly cannot operate *without* technology. Therefore, PSYOPers, with little to no technology, can provide the opportunity to operate in an irregular warfare capacity that can be just as successful as having technological resources.

Irregular Warfare's Place in Cyberspace

Irregular warfare is defined as "a violent struggle among state and non-state actors for legitimacy and influence over the relevant populations" (DoD JP3-13.2 2010, p. xiv). Moreover, Joint Publication 3-13.2 (2010) goes further to state that "irregular warfare is a political struggle for control or influence over, and the support of, a relevant population" (p. xiv). Inevitably, the DoD's engagement in online influence activities, whether offensive or defensive, will move into the realm of irregular warfare. In turn, social media networks can best be understood as unconventional weapons. As such, cyber-based irregular warfare efforts must engage the threat discriminately and apply capabilities indirectly as "technology abhors homogeneity": variation is the standard rather than the exception (Cukier 2005, p. 13).

The conduct of irregular warfare requires studying the confluence of the land, cyber, and human domains. Although social network and link analysis methods are not new to the analytical community, the challenge of collecting the right data, at the right time, and in the right context makes these methods difficult to apply to the irregular warfare fight. If SNA (and influence efforts in general) solely focus on the relationships or direct connections between people, groups, and/or organizations without a thorough understanding of the psychological terrain, any irregular warfare practitioner will be inadequately informed about key aspects of the operating environment (Serrat 2017). Irregular warfare operations require, therefore, an understanding of the political, social, military, economic, terrestrial, and informational architecture of the environment from a broad and deep psychological, social relationship, and cultural perspective. Tarnishing a belligerent's image, disrupting recruitment, countering threat propaganda, building rapport with the local populace, defeating threatening ideologies, and reducing tensions and negative attitudes toward the United States and its allies requires massing effects (Krawchuck 2006). Massing effects requires strategic communication.

To be an effective strategic communicator in the irregular warfare context requires a broad and deep integration of cultural/regional acumen in concert with technical knowledge, skills, abilities, and attributes (KSAAs) (Krawchuck 2006). PSYOP forces are collectively trained and equipped with cultural expertise and KSAAs *before* they use any cyber-based equipment. SNA, and other activities, require certain technical skills, like coding, which CYBERCOM has and PSYOP does not necessarily have. Together, they can work in cross-functional teams with PSYOP in the lead, as the hallmark of the special operations community is a "penchant for creatively incorporating unique and unconventional tools into its arsenal" (Bostick 2011, p. 46). We are in the fourth industrial revolution where there is a hybrid of cyber and physical systems operated by people (Dawson 2020). When using these systems and implementing strategic communication in the irregular warfare environment, "systematic surveys, opinion polls, focus group interviews, and cultural attitudinal databases are just a few examples of tools used to establish baselines of perceptions, monitor social movements, and measure impacts" – all activities that PSYOPers currently do with specific expertise (Krawchuck 2006, p. 38).

Finally, to conduct irregular warfare an influencer must understand the art and science of influence regarding human behavior and its structure and development. The art and science of influence has two key aspects. First, it is rooted in a consistent drive to understand the global information environment from the perspective of all sources of influence including human psychological and social functioning, media, technological, or others (Cook 2014). Second, it is rooted in focusing one's experience, training, and education on leveraging this understanding to initiate actions that change people's attitudes, values, and beliefs, which ultimately underscore and drive behavior (Cook 2014). As essential precursors to any influence campaign, within or outside of the cyber domain, non-kinetic activities and change efforts require an understanding of human behavior in the context of the environment and cross-cultural competence. Arguably, the PSYOPer is their own influence platform. They are a highly effective human weapons delivery system, when appropriately equipped. If influence is the projectile and the PSYOPer is the delivery system, then psychology and human understanding is the gunpowder behind the bullet, be it digital or not.

Massing Effects Requires Synergy, Not Silos

As Nye (2021) plainly articulates, "in cyberspace, one size does not fit all" (p. 42). A key argument to be made here is to avoid a myopic focus on USCYBERCOM versus PSYOP. That will categorically miss the point of where technical expertise and psychological know-how are both needed, as they are finite skills. Effective integration is an imperative, not a "nice to have." To be clear, one cannot replace the other nor specialize in the other's field, which makes cooperation between the two not only ideal but necessary. The capabilities of each community certainly could have temporary priority over each other, yet still have a limited form of mutual dependence (Paul 2014). It is clear that "neither needs the other in order to function, but [each] could be wasted if the other has achieved a desired effect first" (Paul 2014, p. 4). Mutually supporting capabilities can create synergistic effects, achieving things together that neither platform could achieve alone. Scholars and practitioners in the field of influence should not forget the many commands and units (both in the United States and in abroad) with assigned information operations, human intelligence collectors, and other professionals who also have a stake in the influence domain. This is not just about the battlefield of today and tomorrow. Psychological effects are a critical during both war and peacetime (Cowan and Cook 2018).

When thinking of the spectrum of war and peace, or competition and conflict, belief that PSYOP alone can defeat propaganda, malign sentiment, and extremist ideology is a flawed perspective. Ideology is not some fixed entity that infects someone's thinking like a virus, causing consistent behavior (Corman 2014). In fact, ideologies can and do change over time and can be influenced by myriad factors within the environment. Perspectives are something that are social and circulating, influenced by both the individual and their environment, which can change through debate and discourse. If cyberspace is an environmental medium through which discourse, and thusly influence, is happening, then the secret to understanding how malign ideologies produce negative influence – and how to counter that influence – is to recognize the means with which we have not just tactical or technical parity or (hopefully) overmatch, but how communication tools, such as framing and narrative, can be amplified within the cyber domain. That may not require kinetic action, but it will surely require skilled cooperation to achieve non-kinetic effect(s).

There are a few practical ways that the PSYOP and Cyber communities can break out of their silos and work more hand in hand. First, cross-pollination of talent beyond just working within the same problem sets should occur. For example, at a practical level, more PSYOP personnel should be assigned at USCYBERCOM and even at the Army Component Command level, Army Cyber Command. Furthermore, additional PSYOP billets should be included at the highest levels, like the Cyber National Mission Force and Joint Task Force ARES. Conversely, there are no Cyber personnel assigned to the main PSYOP formations at Fort Bragg (4th and 8th PSYOP Groups, respectively). To date, these billets, at low and high levels and in both Cyber forces and PSYOP formations, are nonexistent. Next, creating purpose-built joint task forces for the explicit purposes of combatting a threat is a natural fit, however, creating more established links through regular joint and combined exercises, including official memorandums of agreement and updates to the standing doctrine of both communities, would go a long way to forge the future synergy required to face the realities of war's changing character. Lastly, there should be more connectivity with the schoolhouses for each community. *This is not just about trading best practices and lessons learned* – this would be to share doctrine, shape lesson outcomes, and even share billets for both students and instructors, interlacing the foundations of each side of information warfare.

These and other adjustments would allow U.S. forces to be better positioned to combat the malign narratives, ideological opinions, and extremist sentiments that complicate the battle for the proverbial hearts and minds. Although the United States has proven the capacity to "kinetically engage our enemy at the operational and tactical levels with unsurpassed effectiveness, we have barely begun to take the war to [the enemy] at the strategic level of counter-ideology" (Gorka 2014). The paradox here is that we can be extremely successful at militarily degrading enemies' operational capability, yet our adversaries can become more powerful in the domain of ideological warfare and cyber use. Mass casualty events thusly need to be relooked through the lens of ideology, influence, and media impact. The coin of the realm, therefore, is the narrative, not necessarily weaponry. Bankrupting the coin of the realm will require massed effects and active engagement at the decisive point, shaping the agenda and the story of conflict, and forcing the enemy to lose credibility as much as they lose fighters and supplies.

Conclusion

Of utmost importance is the understanding that actions and activities by the PSYOP community are not meant to replace any component of USCYBERCOM or any other element for that matter. These activities and their specialized training do not give the PSYOPer some "silver bullet" on the battlefield. A concrete understanding of human behavior and an expert competency in foreign cultures clearly differentiate the PSYOPer from the cyber practitioner, but defensive and offensive cyber and a robust understanding of computer systems and other technologies are what cyber specialists also bring to the fight. To wit, Joint Publication 3-13.2 (2010) originally stated that PSYOP actions should be supported by computer network operations. Moreover, computer network operations support PSYOP activities "with dissemination assets to include interactive internet activities [to] scan, deny, or degrade an adversary's ability to access, report, and process information" (p. x). There is space to create synergy, but influence, on the internet or elsewhere, should remain the main effort of the PSYOPer.

Additionally, the use of SNA as dictated by the operational environment (and the information gleaned) should be both shared and deconflicted across *all* available assets. The critical roles that information gathering, analysis, and operations play in the cyber realm to facilitate the timely sharing of cyber threat information can only enhance situational awareness (Grant et al. 2021). The SNA process, targeting cycle, and influence within the cyber domain should supplement all ground-level practitioners with specific tools to capitalize on the exponential increase of the use of the internet as a means of irregular warfare. Disrupting connections within social networks requires more than stopping or infiltrating technology – we must strive to stop digital enemies by changing their desire to weaponize the internet, not just react to the "boom" (e.g., a viral post). Leaders should consider the use of targeting doctrine and tried and true planning processes. Leaders should also consider that as new artificial intelligence systems become more widespread information warfare will also be waged by non-humans.

This chapter asks that tomorrow's practitioners continue to raise the question not only of *who* potentially has the most expertise to conduct information warfare, but also of *what organizations* need to adapt as rapidly as the nature of the battlefield changes. In the future, many problems will not be solved through force alone – and the advent of cyber warfare exacerbates the risk of inadvertent escalation of conflict (Acton 2020). It must be remembered that cyber activities were once seen as the equivalent of nuclear action. Since USCYBERCOM's creation "it has specialized in the conduct of cyber operations and thus has concentrated on acquiring the technical expertise that such operations require" (Lin 2020). Whereas those technical talents are no doubt important, expertise in influence, be it on the internet or otherwise, requires adept psychological understanding of human behavior, how it develops, and how it may be changed.

National security officials as far back as Leon Panetta and Janet Napolitano have warned of a coming "cyber-Pearl Harbor," "cyber-9/11," or "cyber-Armageddon," but it is more useful to normalize the fact that cyberspace will, from here on out, be "a domain of conflict with key terrain that the U.S. will need to take or defend" (Schneider 2021, p. 22). Cyber-enabled information warfare should take advantage of fundamental characteristics of both modern information technology *and* irregular warfare. Certainly, no one could argue against the fact that "cyber operations are intended to hack silicon-based processors and technology, [whereas] psychological operations are intended to hack carbon-based processors (that is, human brains) (Lin 2020)." Ultimately, when building a case for cyber realism, those on the cusp of virtual victory will remind all that geopolitical problems will not have technical solutions (Alperovich 2021). Tomorrow's fight will not be easy – it will require an all-hands-on-deck approach, whether those hands are on a rifle or a keyboard. Technology only magnifies the effects of force employment; however, technology is never a substitute for good force employment, regardless of whether the bullet or bomb is real or digital (Biddle 2004).

References

Acton, J., 2020. Cyber Warfare and Inadvertent Escalation. *Daedulus*, 149(2), 133–149.

Alperovich, D., 2021. The Case for Cyber Realism: Geopolitical Problems Don't Have Technical Solutions. *Foreign Affairs*, 101(1), 44–50.

Amble, J. (Host), 2020. Competition, Conflict, and the Future of Irregular Warfare. [Audio podcast episode]. *In*: *MWI Podcast*, July 22. Modern Warfare Institute. Available from: https://mwi.usma.edu/mwi-podcast-competition-conflict-and-the-future-of-irregular-warfare

Atrews, R., 2020. Cyberwarfare: Threats, Security, Attacks, and Impact. *Journal of Information Warfare*, 19(4), 15–27.

Biddle, S., 2004. *Military Power: Explaining Victory and Defeat in Modern Battle*. Princeton, NJ: Princeton University Press.

Bondyopadhyay, P., 1998. Moore's Law Governs the Silicon Revolution. *Proceedings of the IEEE*, 86, 78–81.

Bostick, R., 2011. Initiating the Cognitive Revolution: An Examination of Special Operations Military Information Support Operations. *U.S. Army War College*, 1–29.

Bourne, K., 2019. Targeting in Multi-domain Operations. *Military Review*, 99(3), 60–67.

Brown, J., 2012. Improving Nonlethal Targeting: A Social Network Analysis for Military Planners [Master's Thesis, Naval Postgraduate School, Monterey, CA]. Dudley Know Library, Naval Postgraduate School, 1–27. https://hdl.handle.net/10945/27800. Available from: https://calhoun.nps.edu/bitstream/handle/10945/27800/12Dec_Brown_Jason.pdf

Cilluffo, F. and Clark, J., 2014. Repurposing Cyber Command. *Parameters*, 43(4), 111–118.

Cook, C., 2014. Continuing Education: The Brains Behind the Brawn of the Operator. *IO Sphere*, Winter 2014, 22–25.

Cook, C., 2022. PSYOP, Cyber, and Internet Influence: Firing Digital Bullets. *Journal of Information Warfare*, 21(2). Available from: www.jinfowar.com/journal/volume-21-issue-2/psyop-cyber-internet-influence-firing-digital-bullets

Cook, C., and Collins, L., 2021. PSYOP, Cyber, and Infowar: Combatting the New Age IED. *Modern Warfare Institute*, April 6. Available from: https://mwi.usma.edu/psyop-cyber-and-infowar-combating-the-new-age-ied/

Corman, 2014., Ideology, Framing, and Narrative. *Influence*, 1(4), 19–22.

Cowan, D., and Cook, C., 2018. What's in a Name? Psychological Operations versus Military Information Support Operations and an Analysis of Organizational Change. *Military Review*, March 2018, 1–7.

Cukier, K., 2005. Who Will Control the Internet? *Foreign Affairs*, November/December, 7–13.

Czege, H., 2020. *Commentary on the US Army in Multi-Domain Operations 2028*. Carlisle, PA: Strategic Studies Institute and US Army War College Press, 1–66.

Dawson, M., 2020. *Cyber Warfare Threats and Opportunities*. Thesis (PhD). Universidade Fernando Pessoa.

Duggan, P., 2014. UW in Cyberspace: The Cyber-UW Pilot Team Concept. *Special Warfare*, 27(1), 68–70.

Federal Bureau of Investigation (FBI), 2018. The "Morris Worm": 30 Years Since First Major Attack on the Internet. *FBI*, 2 November. Available from: www.fbi.gov/news/stories/morris-worm-30-years-since-first-major-attack-on-internet-110218

Gendon, G., Blass-Irizarry, H., and Boggs, J., 2009. Next Generation Strategic Communication: Building Influence Through Online Social Networking. *Joint Forces Staff College*, 1–18.

Gjelten, T., 2013. First Strike: U.S. Cyber Warriors Seize the Offensive. *World Affairs*, 175(5), 33–43.

Gordon, S. and Rosenbach, E., 2021. America's Cyber-Reckoning: How to Fix a Failing Strategy. *Foreign Affairs*, 101(1), 10–20.

Gorka, S., 2014. The Importance of Ideology in Special Warfare. *Influence*, 1(3), 11–13.

Grant, A., Billman, A., Cell, T., Meador, B., Halter, T., Hartley-McBride, S., and Kaspar, B., 2021. Critical Roles of Information, Analysis, Research, and Operations in the Cyber Realm. *Journal of Information Warfare*, 20(2), 67–80.

Headquarters, Department of the United States Army, 2013. *Army Field Manual Number 3-53: Military Information Support Operations*. Washington, DC: U.S. Government Printing Office.

Krawchuck, F., 2006. Strategic Communication: An Integral Component of Counterinsurgency Operations. *Connections: The Quarterly Journal*, 46, 35–50.

Lin, H., 2020. On the Integration of Psychological Operations with Cyber Operations. *Lawfare*, 9 October. Available from: www.lawfareblog.com/integration-psychological-operations-cyber-operations

Lin, H. and Kerr, J., 2021. On Cyber-Enabled Information Warfare and Information Operations. *In*: *Oxford Handbook of Cybersecurity*. New York: Oxford University Press, 1–29.

Moore, G., 1965. Cramming More Component onto Integrated Circuits. *Electronics*, 8, 114–117.

Naim, M., 2014. *The End of Power*. New York, NY: Basic Books.

Nye, J., 2021. The End of Cyber-Anarchy? How to Build a New Digital Order. *Foreign Affairs*, *101*(1), 32–42.

Paul, C., 2014. Integrating Apples, Oranges, Pianos, Volkswagens, and Skyscrapers: On the Relationships Between Information-Related Capabilities and Other Lines of Operation. *IO Sphere*, Winter 2014, 3–5.

Paul, C., 2020. Understanding and Pursuing Information Advantage. *The Cyber Defense Review*, Summer 2020, 109–123.

Rid, T. and Hecker, M., 2009. *War 2.0: Irregular Warfare in the Information Age*. Westport, CT: Praeger.

Rouse, E., 2021. *PSYOP Creed*. *Psywarrior*, 10 December. Available from: www.psywarrior.com/creed.html

Schneider, J., 2021. A World Without Trust: The Insidious Cyber Threat. *Foreign Affairs*, 101(1), 22–31.

Serrat, O., 2017. Social Network Analysis. *In*: O. Serrat, ed. *Knowledge Solutions*. Singapore: Springer, 39–43.

Shallcross, N., 2017. Social Media and Information Operations in the 21st Century. *Journal of Information Warfare*, 16(1), 1–12.

Shirky, C., 2011. The political power of social media. *Foreign Affairs*, *90*(1), 28–41.

Staton, W., 2020. A Millennial's Perspective on the Legacy of Vietnam. *Medium*, 21 October. Available from: https://medium.com/@WStaton85/a-millennial-s-perspective-on-the-legacy-of-vietnam-21e247dde019

Turnley, J., 2011. Cross-Cultural Competence and Small Groups: Why SOF are The Way SOF Are. *Joint Special Operations University Report*, 11–1.

United States Cyber Command, 2021. Our History. *US Cyber Command*, 23 August. Available from: www.cybercom.mil/About/History/

United States Department of Defense (DoD), 2010. *Joint publication 3-13.2: Psychological Operations*. Available from: https://fas.org/irp/doddir/dod/jp3-13-2.pdf

United States Department of Defense (DoD), 2014a. *Joint Publication 1-02: Department of Defense Dictionary of Military and Associated Terms*. Available from: www.dtic.mil/doctrine/new_pubs/jp1_02.pdf

United States Department of Defense (DoD), 2014b. *Joint Publication 3-13: Information Operations*. Available from: www.jcs.mil/Portals/36/Documents/Doctrine/pubs/jp3_13.pdf

United States Department of Defense (DoD), 2015a. *Field Manual 3-05.301: Psychological Operations, Tactics, Techniques, and Procedures*. Available from: https://fas.org/irp/doddir/army/fm3-05-301.pdf

United States Department of Defense (DoD), 2015b. *Field Manual 3-60: The Targeting Process*. Available from: https://armypubs.army.mil/epubs/DR_pubs/DR_a/pdf/web/atp3_60.pdf

Ventre, D., 2016. *Information Warfare*. Hoboken, NJ: Wiley.
von Clausewitz, C., 1984. *On War*. trans. Howard, M., Paret, P., and Brodie, B. Princeton, NJ: Princeton University Press.
Widener, C., 2008. *The Art of Influence: Persuading Others Begins with You*. New York: Crown Business.

10 Cybersecurity as a Public Good

Government Intervention Is Only Part of the Solution

Margaret W. Smith and Jim Monken

Introduction: Ransomware, An Equal Opportunity Threat

Ransomware is an equal-opportunity threat and all organizations are potential targets (Hudak 2021). In the broader context of great power competition (GPC), open societies, like the United States (U.S.), are increasingly at risk and the safe harbor that ransomware and other cybercriminals are tacitly (or overtly) granted by the likes of North Korea, Iran, China, and Russia, provide freedom of movement to malicious cyber actors with a low threat of meaningful consequences (Sanger and Perlroth 2021). Recent history alone provides ample evidence of ransomware's nondiscriminatory nature, with hacks ranging from Ireland's national health service to Costa Rican government websites to Dole Food Company (Perlroth and Satariano 2021; Reed 2022; Popovici 2023). Because everything that touches the internet is vulnerable – from gasoline to meat, or even a ferry service – it signals a cyberspace that is increasingly hostile to law-abiding and legitimate users (Rosenbaum 2021). And the problem is only getting worse – each instance of ransomware and cybercrime elevates the risk of future attacks in the public psyche and, with many top experts calling the U.S. Government's response inadequate, the question of how to stem attacks and reduce their impact remains elusive (DARK Reading 2021).

With cyberspace as the vehicle or medium by which ransomware and malicious code are delivered, hacks are often viewed as a technical problem necessitating a technical solution. However, the focus on technical solutions means that the policy and economic aspects of the ransomware and cybercrime problem remain underdeveloped despite the recent convening of 30 countries to launch a U.S. led global initiative to combat cybercrime (Gatlan 2021). Ultimately, the rapid convergence of information technology (IT) and operations technology (OT) in the systems that manage and operate our critical infrastructure – from electric grids to pipelines to financial trading platforms – indicates that risk mitigation requires a comparable merging of strategy and policy to meaningfully lessen the impact a cyberattack can have on software-based and physical systems and the customers those systems serve (i-Scoop n.d.). Achieving a robust, multi-sector cybersecurity posture requires a new level of coordination within the cyber ecosystem with clearly delineated roles,

DOI: 10.4324/9781003425304-11

responsibilities, tactics, and services to maximize technical capabilities and jurisdictional authority to mitigate risk (Smith and Monken 2021).

Why Ransomware? It's Profitable!

Criminal entities that utilize ransomware have decidedly economic objectives – in lieu of political objectives – when executing their attacks despite the frequent connections made between ransomware groups and nation-state adversaries (Reuters 2021). Ransom payments are, on average, becoming larger: Coveware (2019, 2023) has reported that average payments have increased from roughly $6,733 per attack at the end of 2018 to over $327,883 per attack in mid-2023, which means profitability for criminal actors is increasing too. Additionally, the rise of ransomware-as-a-service (RaaS) – a variant of the software-as-a-service business model – means that ransomware, and its use, are not limited to those who develop the capability. Selling pre-developed ransomware to criminals to use for extortion is becoming a frequent practice and decreases the risk faced by ransomware developers since they are not the ones executing the attacks (Midler 2020). For criminal organizations that purchase ransomware, RaaS reduces costs since malicious actors can purchase prebuilt ransomware instead of expending the resources and time to develop it in-house. Therefore, RaaS effectively "expands the ransomware threat landscape" because criminal entities can outsource their ransomware development or purchase an on-demand service (Midler 2020).

Because profit is the primary objective, prevention efforts should focus on changing the incentive structure of the transaction and reducing the incentives of using ransomware to the point where hacking leaves the threat actor worse off, while also mitigating the effect on the U.S. consumer and preventing market shocks. Technical solutions alone cannot generate the multifaceted changes needed to affect the ransomware business and make it unappealing to criminals. For example, the RaaS entity, Darkside, quickly recognized that their successful hack of Colonial Pipeline resulted in an increased demand for gas along the East Coast – a social cost that stressed the market and consumers (Schwirtz and Perlroth 2021; Myre 2021; Krauss and Sanger 2021). And because of that social cost, the hack drew substantial amounts of unwanted attention to the Darkside criminal outfit, provoking a coordinated government response that seized stolen funds and servers, and will result in criminal charges, thereby limiting the actors' freedom of movement (Dailey 2021).

In fact, the Colonial Pipeline ransomware attack attracted so much public and media attention that Darkside allegedly shut down its operations after losing access to its public-facing infrastructure and private servers (Krebs 2021). Other criminal groups also pushed their ransomware operations further underground or abandoned RaaS operations altogether (although this was temporary). Cybercrime forums at the time indicated that RaaS outfits writ large felt the impact of Darkside's hack, with groups claiming that their infrastructure was also taken offline, that they were amending the rules by which they operate, or that they had abandoned the ransomware business altogether because of the large amount of negative attention directed at them (Intel 471 2021). Therefore, there is evidence that the multi-sector response to Darkside's hack temporarily altered ransomware transaction costs for criminals: the operational context and market incentives changed, and, as a result, ransomware suddenly became much more costly.

Cybersecurity Is Not a Pure Public Good: The Limits of Government Intervention

Even though cybersecurity has many characteristics of a public good – a commodity or service that is made available to all members of society – most information security and physical security decisions are made by individual stakeholders for individual interests and within individual resource constraints. Back in 2003, former Department of Homeland Security Secretary, Tom Ridge (2003), recognized that

> Anywhere there is a computer ... whether in a corporate building, a home office, or a dorm room ... if that computer isn't secure, it represents a weak link. Because it only takes one vulnerable system to start a chain reaction that can lead to devastating results.

Taking Ridge's comment literally, the U.S. government cannot correct the public good problem of cybersecurity alone – especially because corporate and private interests are the driving factor of many cybersecurity decisions (or indecisions).

A major hurdle to securing critical systems is that the optimal level of cybersecurity is not static, but dynamic and is not universal (e.g., organization A's security needs are different than organization B's needs which are different from user C's needs, etc., and, unless organizations are completely siloed, low security in organization A may provide an adversary access to Organization C), making it impossible for the government to deliver a single solution to meet all users' needs. Moreover, the private sector already devotes extensive resources to cybersecurity because private industry has a lot at stake – a recent McKinsey report found that organizations around the world spent around $150 billion in 2021 on cybersecurity (Aiyer et al. 2022). Therefore, most of the private sector is already providing the positive externality – a cost or benefit that is borne by a stakeholder who did not agree to the action that caused the cost or benefit – of robust cybersecurity (e.g., securing consumer assets, data, and information) for self-interested reasons, and a new subsidy or tax cut is unlikely to drastically change or incentivize an organization to do more (Powell 2005). Lastly, because cyberspace is borderless and trillions of devices are interconnected in a mix of private and public networks, decentralized cybersecurity decisions are plagued by externalities that often result in derisory security levels that put entire systems at risk.

In many ways, the markets for RaaS and the security measures to prevent ransomware attacks are evolving together. Co-evolution makes it difficult to devise effective cybersecurity solutions because the system is so complex – the global domain of cyberspace, the interdependence of stakeholders, and the heterogeneity of players all add to the problem's complexity. And yet, the Colonial Pipeline hack is an example of how a coordinated response, one that makes use of government assets and private sector cooperation, can impose costs on cybercriminals and potentially alter the incentive structure in the ransomware market altogether. Importantly, the multifaceted response to Darkside highlights the areas where collaboration on information infrastructure can generate many security-enhancing incentives, and where negative externalities (e.g., critical service disruptions) resulting from sub-optimal or lax cybersecurity exist that may require voluntary or government-led collective measures to correct (e.g., legislation or regulations).

The Realities of Ransomware and Cybercrime: There's No Silver Bullet

After the Colonial Pipeline hack, the U.S. Department of Justice (DoJ) announced that it had seized a substantial portion of the ransom – 63.7 Bitcoins, valued at over $2.3 million – and broke the myth that Bitcoin was untraceable (Benner and Perlroth 2021; Perlroth et al. 2021). The result was a change to the context in which cybercriminals operate, a shakeup of transaction costs, and a chilling of the RaaS market. However, the ransomware problem – and the problem of cybercrime more generally – is more complex than simply evaluating transaction costs related to the characteristics of cryptocurrency.

First, Bitcoin operates on a public or open ledger system, meaning that transfers are not anonymous or obfuscated – anyone reading the ledger can see exactly the time, amount, and parties involved by their wallet numbers. Importantly, even though Bitcoin was recovered in the Colonial Pipeline case, it is *not* recoverable 100% of the time. The Bitcoin in question must first be in a law enforcement entity's jurisdiction, and it must be "seize-able" – i.e., able to be taken or not yet converted into another store of value (e.g., fiat). Also, to punish someone for stealing Bitcoin, law enforcement must be able to connect the Bitcoin wallet that holds the stolen currency to an actor. Other cryptocurrency assets not operating on an open ledger system are available, but Bitcoin is largely adopted because of its value – the other closed systems are not yet as valuable as Bitcoin but may be much more difficult for law enforcement to seize.

Furthermore, we know the ransomware problem is escalating, but it is difficult to assess the degree of escalation and the actual impact ransomware has on national security. In a recent survey of businesses by Cloudwards, 50% of companies reported being the victim of at least one ransomware attack, and trends also show that ransomware demands are increasing, too (Kochovski 2022). Both factors – the lack of reporting and increasing ransoms – mean that any company attacked immediately faces two related questions: should we pay the ransom in exchange for the decryption key? And do we report the hack and risk damage to our brand and reputation? Since ransomware was first developed, many victims have quietly opted to pay – a legal and tax-deductible option for companies, but one that the Federal Bureau of Investigation (FBI n. d.) discourages – figuring the ransom payment to be cheaper than the cost of rebuilding systems, data, services, and consumer trust.

Unfortunately, paying *is* cheaper – and quieter – in most cases and, for most companies and organizations, the only alternative to paying for a decryption key is spending substantially *more* resources and time to rebuild systems from scratch – as the Baltimore County Public Schools continue to discover after the district was hit by ransomware in November 2020 (Simpson 2021).

A counter-factual question is also important to note when considering the Cloudwards survey: how many ransomware attacks go unreported? Companies are often compelled to report a cyber security incident, like a data breach, to regulators and authorities but reporting on ransomware attacks was not required until recently. In March 2022, it became mandatory in the U.S. for covered entities in any of the 16 critical infrastructure sectors to report cybersecurity incidents and any payments within specified time frames to the Cybersecurity and Infrastructure Security Agency (CISA) (CISA 2023). The Cyber Incident Reporting for Critical Infrastructure Act of 2022 (CIRCIA) is buried in the Consolidated Appropriations Act of 2022 on pages 2,542–2,581 and depends on several factors: the extent of the breach (e.g., how many records were affected), the type of data exposed or stolen, where the breach took place (e.g., certain laws only apply to certain geographic locations/countries), and the industry or sector the breach occurred in (e.g., industry-specific rules vary on data breach notification). Though they are not fully clear, page 2,543 of CIRCIA does establish some guidelines for what a substantial cyber incident may look like:

- Occurrences of substantial loss to confidentiality, integrity, and availability of information systems, or serious impact to safety and resiliency of operational systems and processes.
- Disruption of business or industrial operations.
- Unauthorized access or disruption of business or industrial operations caused by third parties.
- The number of people impacted.
- Impacts on industrial control systems (Platsis 2022).

CIRCIA also states that reporting entities are entitled to the protections against liability described in section 106 of the Cybersecurity Act of 2015, however, the law remains open for public comments, as CISA is charged with finalizing the rules and definitions under consideration, so the result is far from determined.

Alternatively, for companies in Europe, and those international companies doing business in Europe, they are legally obligated under the General Data Protection Regulation (GDPR) to inform the Information Commissioner's Office if they suffer a breach that involves the personal information of customers or employees (Nadeau 2020). Similar obligations also exist in the U.S. for healthcare data protected under the Health Insurance Portability and Accountability Act of 1996. But until CIRCIA became law in the U.S. ransomware was considered different and various reporting obligations existed for victims of cybercrime, which made tracking and assessing the true extent of the costs and victims difficult. In 2018, for example, the FBI's Internal Crime Complaint Center reported over 350,000 cybercrimes took place but also estimated that *only 15% of victims reported* their crimes to a law enforcement entity (FBI 2019, FBI 2016). As mentioned above, in 2018, the average attack imposed a ransom of roughly $6,733 and, despite the large jump to $327,883 in 2023, the cost of paying off the criminals is not crushing to a large company's bottom line.

Reporting even a $327,883 ransomware attack will likely generate a higher cost for companies than simply paying the fee and moving on with operations. Small, distributed attacks are therefore incentivized under this model – if the benefits of reporting are minimal and the costs low, why would a company report a cybercrime? Ultimately, companies are not incentivized to report cybercrime when the financial impact is small and remain unlikely to do so if they do not fall within one of the 16 critical infrastructure sectors.

Yet, even the rollouts of mandatory technology standards, as advised in the May 12, 2021, Executive Order (EO) 14028 on "Improving the Nation's Cybersecurity," and mandatory reporting criteria, as outlined in CIRCIA, are multiyear, multibillion-dollar propositions, as the electricity industry demonstrated over its two-decade transition from voluntary to mandatory standards under the watchful eye of the Federal Energy Regulatory Commission and the North American Electricity Reliability Corporation (Biden 2021). Even today, critical infrastructure protection standards are not considered synonymous with totally secure systems; rather, they represent the minimum of what is expected and rely on the assurance that cybersecurity programs are subject to recurring checks on quality and completeness. To truly compete in the era of GPC, the relationship between government and the private sector can no longer be limited to a "carrots and sticks" approach of rates, regulation, and reporting but needs to become a genuine partnership capable of integrating currently disparate intelligence programs, conducting joint cyber defensive operations, and collaborating on threat identification and mitigation strategies.

What Can Be Done?

Even though substantial investments in cybersecurity are made each year – by private and public entities alike – the annual number of cyberattacks on companies and public entities has not decreased, and larger investments in cybersecurity do not necessarily translate into a higher cybersecurity maturity level (Pomerleau and Maimon 2020; Deloitte 2020). Much of the problem lies in the nature of cybercrime and cyberattacks themselves – human beings commit them, take advantage of peoples' and organizations' vulnerabilities, and are not wholly separated from the physical world (Pomerleau and Maimon 2020). As much as we would like to blame computers and interconnectedness for the increasing levels of malicious cyber activity, technology often plays an incidental role, and instead, it is the complex interactions between humans and technology that generate opportunities for criminals and adversaries to exploit.

Recently, as the SolarWinds hack was just becoming public knowledge, Jill Lepore wrote an ominous article for the *New Yorker* titled, "The Next Cyber Attack is Already Underway" (2021). While the title was likely intended to raise eyebrows, it is nonetheless true – attacks and hacks are ongoing, evolving, and continuous; and it will always be that way. As the Colonial Pipeline hack showed, IT and OT systems converge, creating new challenges for physical security and cybersecurity. Because cyberspace remains a land of opportunity, criminals and nation-states will continue to take advantage of outdated software, unpatched systems, and static defenses. Ultimately, to be good at cybersecurity means accepting that defenses in place today may be obsolete tomorrow and assuming that your systems are already breached.

Lepore (2021) also adds that "the federal government is effectively insecure." And while security is certainly a moving target in cyberspace, EO 14028 and CIRCIA give the U.S. the opportunity to move toward creating a federal cybersecurity eco-system that fosters information sharing and enables CISA to take a more dynamic approach to defending the .gov domain while also partnering with the private sector to protect the .com domain. In the absence of a cybersecurity ecosystem rooted in a multisector and multidisciplinary approach, customers ranging from individuals gassing up their cars to the Department of Defense – which consumed approximately *eighty-eight million barrels* of fuel in 2020, or roughly 3.7 billion gallons – remain exposed to a common threat (U.S. Department of Defense 2020; Baskin and Gessler 2020). Therefore, to improve the U.S.'s cybersecurity posture, our approach should be just as complex and multifaceted as cyberspace itself.

Incentivize Critical Service Continuity

First, it is essential to recognize that the Colonial Pipeline attack showed how a disruption in a company's IT (business and accounting) system could impact its OT (production and distribution) systems by compromising the company's ability to accurately charge and bill consumers for their product (Smith and Monken 2021). Company interests (e.g., accurate billing and payment) collided with public interests (e.g., gasoline supply and cost) and company interests prevailed – Colonial Pipeline was incentivized to shut down its OT systems to avoid insolvency and, as a result, consumers experienced increased demand, decreased production, public panic, and rising costs even though Colonial's OT systems were *unaffected* by the original hack. Ultimately, if the incentive structure were different, gas *could* have kept flowing along the Eastern Seaboard without disruption.

To change the operational context of critical service delivery, the government should establish a Federal Commodity Disruption Fund (FCDF) to incentivize companies to avoid service curtailments that are mostly driven by the temporary risk of financial harm in the wake of a cyberattack. Access to FCDF should be conditional and based on a contractual agreement to notify and involve federal law enforcement upon hack discovery, a documented ability to rapidly recover business systems and operational assets, and a commitment to fully reimburse the funds after normal operations resume. As a proverbial carrot for companies to con-tinue delivering critical services, FCDF eligibility should also hinge on a company maintaining baseline standards for cybersecurity – a recommendation that gained many advocates following the recent wave of attacks on critical sectors that currently operate in a largely unregulated or wholly voluntary cybersecurity environment (Patel 2021).

The FCDF should be funded through the appropriations process as part of CISA's budget, like the Disaster Relief Fund (DRF) under FEMA. However, unlike DRF funding, FCDF is short-term, targeted and must be repaid – it is meant as gap coverage for commodity and critical service continuity, not complete recovery. As such, FCDF cannot work alone and should instead be considered one response mechanism in the cybersecurity ecosystem. As a type of gap coverage or bridge loan, FCDF funding helps to underwrite some of the risk associated with digital provision of critical services – attacks are inevitable but do not have to be socially costly in the short term. Finally, cybersecurity standards should also expand on the current, advanced persistent threat (APT)-driven security guidelines. At a minimum, baseline FCDF compliance should include guidelines for systems and procedures tailored to ransomware (e.g., two-factor authentication, network segmentation, data back-ups, machine replacement "throughput," etc.) as well as ensuring those guidelines are updated on an ongoing and consistent basis to account for new attacks (e.g., CISA-FBI guidance for the Kaseya VSA supply-chain ransomware attack) (CISA 2021).

Incentivize Risk Management and Mitigation

Cybersecurity risk is extremely difficult to assess – partly because the historical data on cybersecurity incidents is inconsistent and incomplete and, because the second and third-order effects of a cybersecurity incident are hard to capture or measure due to interdependencies. Companies and governments not only face threats to their own systems, but also face systemic risk: if Company A depends upon the computerized systems of Company B (e.g., a computerized logistics systems at a transit hub to move goods around), then Company A is linked to Company B's systems' security. For international corporations, system interdependencies can span the globe and industries, making systemic risk enormous and insurmountable. Despite these challenges, the cyber insurance market has grown rapidly since the American Insurance Group offered the first insurance policy for cybersecurity in 1997, and demand for insurance has increased alongside risk in cyberspace – companies and organizations are purchasing cyber insurance to underwrite that risk and to help cover the cost of a catastrophic cyber event, and Congress has even recognized the importance of a stable insurance market to the overall cybersecurity ecosystem (Brown 2014; Crain Communications 2021; Government Accountability Office 2021).

To investigate the current market, the National Defense Authorization Act (NDAA) for Fiscal Year 2021 included a provision for the Government Accountability Office (GAO) to study the U.S. cyber insurance market to better understand the market's role in offsetting the costs of responding to and recovering from cyberattacks (House Armed Services Committee 2022; GAO 2021). GAO analyzed industry data on cyber insurance policies and reviewed reports on cyber risk and cyber insurance from researchers, think tanks, and the insurance industry and found that more insurance clients are opting for cyber coverage. What GAO observed is that insurers are *lowering* coverage limits while *raising* premiums for high-risk sectors. As mentioned above, one of the problems the cyber insurance market faces is that historical data on cyber-attacks is not comprehensive and is, in many cases, low-quality (i.e., incomplete, vague, and/or missing completely), making it difficult to appropriately price insurance policies – a problem that is likely driving the lower coverage limits and higher premiums. The current incentive structure does not encourage companies to report a cyber incident, which further exasperates the challenges inherent to cyber risk assessments makes it difficult to determine proper coverage.

Unlike a natural disaster, a cybersecurity incident is viewed as preventable, and companies are expected to take responsible measures to protect systems, data, and privacy. The ransomware crisis has placed significant strain on the cyber insurance industry by increasing the volume (or frequency) of claims and their value. As a result, premiums are rising and the threshold for accessing insurance is rising too – with more claims and more risk, cyber insurance brokers are finding it necessary to turn clients away if they do not meet increasingly costly baseline levels cybersecurity. Simultaneously, cyber insurance is quickly becoming a requirement to do business in several sectors. Combined, the threats to the cyber insurance industry are having counter-balancing effects: businesses are making additional cybersecurity investments to ensure their systems are up-to-date and protected and, more organizations are finding themselves priced out of cyber insurance and facing insurmountable risk (Glover 2022).

Reporting Standards and Requirements

Underreporting of cybercrime matters for a variety of reasons. For example, malicious cyber actors are more inclined to target companies in sectors or industries where reporting is less likely. Brand-oriented and trust-based companies managing large customer assets or sensitive information are likely to be more concerned about losing customer trust and the damage to their corporate brand, making them a more attractive target for cybercriminals. Underreporting also makes it difficult to accurately assess the extent of the ransomware problem or to infer trends of which industries are more at risk of attack. Another symptom of underreporting is a deep-seated notion that reporting is futile, leaving many organizations to wonder, "What's the point?" Determining attribution and going after cybercriminals (especially foreign actors) is difficult, and, even when identified, law enforcement is ill-equipped to restore operations and recover stolen data or assets before they change hands. Therefore, companies often opt to simply pay the ransom to restore operations quickly and to keep security lapses out of the public domain.

Because ransomware's impact extends well beyond the victim's balance sheet and has implications for national security in the aggregate, the decision to pay or not is a complicated one. While JBS paid its attackers $11 million in ransom, the attack also disrupted roughly 10,000 jobs, and the "price for a US boxed beef cutout went up by $3/cwt (a standard unit of weight) for Choice [grade], and 5.55 cents for Select grade on June 1 compared with May 28" – a tidy sum in a $1.2 trillion global industry and consumers (Robbins 2021; Dutton 2021; Business Wire 2020). DarkSide's hack also impacted markets but, the hack also highlights how Colonial Pipeline's early cooperation with the FBI was a major factor in the successful recovery effort, with federal officials even going so far as to credit "the company for its role in a first-of-its-kind effort by a new ransomware task force" within the DoJ (Benner and Perlroth 2021). The Colonial Pipeline case highlights how, when corporate and public interests collide in the aftermath of an attack, there is a role for government, and intervention is needed to ensure better outcomes for the victim and consumers alike.

A System of Systems Approach

Managing costs is key to developing systems' robustness for our cybersecurity eco-system. Costs are not simply financial – time, knowledge, access to services, and other non-tangible costs, many of which do come down to money – are also factors to consider. Additionally, the collective responsibility for a cybersecurity eco-system does not mean that the systems' robustness needs to come free of charge, at equal cost to all users, or that all users need the same exact security or protection to determine access. Ultimately, the federal government's role is to ensure that all users have access to digital technologies adequate to the purpose and the context of their deployment and employment in specific and unique use cases within the broader ecosystem.

However, to be robust, our cybersecurity ecosystem will require governments – at all levels – to shoulder some of the costs, including those related to standards setting, certification procedures, testing and verification technologies, and more. But managing our cybersecurity ecosystem is important for many reasons, but namely, government intervention and responsibility will enable a systematic approach and foster a collective sense of responsibility among the different stakeholders and across sectors while also encouraging collaboration. To be effective, direct and indirect externalities must be considered, and medium and long-term consequences should be factored into standards and regulations developed to improve our col-lective security posture. Our cybersecurity ecosystem is rife with interdepend-encies and relies upon the security of distinct but connected technologies, their impact on the security context of their use, and the public interest.

Management of the cybersecurity ecosystem mandates collaboration between the public and private sectors to ensure that systems' robustness can meet the needs of public interest – private interests, like accurate billing in the Colonial Pipeline case, that pose a risk to public services and goods if attacked, should be incentivized to align with the public interest. Additionally, the government must take the lead and set standards, certification requirements, and testing and verifi-cation procedures to ensure the minimum level of cybersecurity required to sus-tain the ecosystem is maintained and regulated. Simultaneously, the private sector has the responsibility to design, with cybersecurity in mind, building systems and processes that improve upon cybersecurity methods approved for their industry or sector. Additionally, the public and private sectors should collaborate around developing controlling and testing mechanisms suitable for different industries.

Moreover, to be effective, there is first a need for increased dependency mapping of IT/OT applications and data to inform risk analysis and improve our understanding of where critical systems are most vulnerable. At the same time, the public sector can support dependency mapping by including information-sharing and collaboration as part of its capability-building initiatives and procedures. The distribution of responsibilities among the various stakeholders, together with the need to consider direct and indirect externalities, is likely to foster collaboration and information sharing. A collective understanding of the vulnerabilities inherent to different systems involved in the same supply chain or data supply chain, for example, is necessary for the private sector to prove systems' robustness and to align security priorities across all systems involved.

The collaboration will generate a deterrent effect. With a private sector that aligns itself with the broader public goals of the cybersecurity ecosystem, retaliation is possible. The impact to East Coast gas production prompted a public outcry, but because authorities got involved early in the investigation, the FBI was able to conduct a timely investigation, and ultimately, most of the ransomed funds were mostly retrieved. The information learned from the attack implies that a rapid joint response effort is effective. In the weeks and months immediately following the attack on Colonial Pipeline, the RaaS outfit, Darkside went dark, showing how the fanfare and news media attention focused on the group changed the context and incentive structure of the broader RaaS market. In the future, a similar chain of events (see Figure 10.1) should be used to respond to ransomware attacks.

Finally, to better define roles and responsibilities within the cybersecurity ecosystem – including those for the private and public sectors, state and local governments, and the public – we need to stop thinking about cybersecurity as a public good that needs to be provided and instead focus on the collective responsibility – shared by corporations, governments, and private citizens – of creating a robust cybersecurity ecosystem. Government provision and standards are only one part of a multi-solution approach required that makes cybersecurity different than the oft-cited public good of national security. Instead, as FBI Director, Christopher Wray recently explained, for cybersecurity, "[t]here's a shared responsibility, not

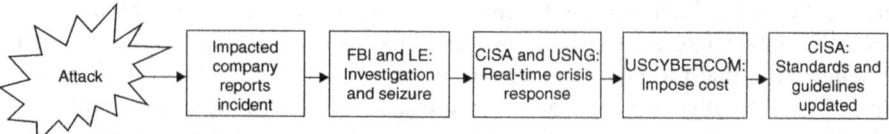

Figure 10.1 Response Flow in the U.S. Cybersecurity Ecosystem.

Source: Author's Creation.

just across government agencies, but across the private sector and even the average American" to engage with the cybersecurity ecosystem (Barnes 2021).

References

Aiyer, B., et al., 2022. New Survey Reveals $2 Trillion Market Opportunity for Cybersecurity Technology and Service Providers. *Risk and Resilience, McKinsey*, 27 October. Available from: www.mckinsey.com/capabilities/risk-and-resilience/our-insights/cybersecurity/new-survey-reveals-2-trillion-dollar-market-opportunity-for-cybersecurity-technology-and-service-providers

Barnes, J. E., 2021. F.B.I. Director Compares Danger of Ransomware to 9/11 Terror Threat. *The New York Times*, 4 June. Available from: www.nytimes.com/2021/06/04/us/politics/ransomware-cyberattacks-sept-11-fbi.html

Baskin, M. and Gessler, K., 2020. How Plummeting Fuel Prices and Reduced Operations Could Free Up Billions of Defense Dollars. *War on the Rocks*, 26 May. Available from: https://warontherocks.com/2020/05/how-plummeting-fuel-prices-and-reduced-operations-could-free-up-billions-of-defense-dollars/

Benner, K. and Perlroth, N., 2021. U.S. Seizes Share of Ransom from Hackers in Colonial Pipeline Attack. *The New York Times*, 7 June. Available from: www.nytimes.com/2021/06/07/us/politics/pipeline-attack.html

Biden, J. R., 2021. Executive Order 14028: Improving the Nation's Cybersecurity. *Code of Federal Regulations*, 12 May. Available from: www.whitehouse.gov/briefing-room/presidential-actions/2021/05/12/executive-order-on-improving-the-nations-cybersecurity/

Brown, B. D., 2014. The Ever-Evolving Nature of Cyber Coverage. *Insurance Journal*, 22 September. Available from: www.insurancejournal.com/magazines/mag-features/2014/09/22/340633.htm

Business Wire, 2020. Global Meat Industry Almanac 2020: Market Value and Volume 2015–2019 and Forecast to 2024. *BusinessWire*, 20 December. Available from: www.businesswire.com/news/home/20201224005114/en/Global-Meat-Industry-Almanac-2020-Market-Value-and-Volume-2015-2019-and-Forecast-to-2024---ResearchAndMarkets.com

Consolidated Appropriations Act of 2015, Public Law 113, 114th Cong., 2015, 18 December. Available from: www.govinfo.gov/content/pkg/PLAW-114publ113/pdf/PLAW-114publ113.pdf

Coveware, 2019. Coveware's 2018 Q4 Ransomware Marketplace Report. *Ransomware Recovery Blog*, 22 January. Available from: www.coveware.com/blog/2019/1/21/coveware-2018-q4-ransomware-marketplace-report

Coveware, 2023. Big Game Hunting is back despite decreasing Ransom Payment Amounts. *Ransomware Recovery Blog*, 28 April. Available from: www.coveware.com/blog/2023/4/28/big-game-hunting-is-back-despite-decreasing-ransom-payment-amounts

Crain Communications, 2021. Why Businesses Are Paying More for Cyber Insurance. *Crain's Daily Gist Podcast*, 15 June. Available from: www.chicagobusiness.com/crains-daily-gist/crains-daily-gist-podcast-why-businesses-are-paying-more-cyber-insurance?

Cyber Incident Reporting for Critical Infrastructure Act of 2022, Public Law 103, 117th Cong., 2022, 15 March, 2,542–81. Available from: www.congress.gov/bill/117th-congress/house-bill/2471/text

Cybersecurity and Infrastructure Security Agency, 2021. CISA-FBI Guidance for MSPs and Their Customers Affected by the Kaseya VSA Supply-Chain Ransomware Attack.

National Cyber Awareness System, 6 July. Available from: https://us-cert.cisa.gov/ncas/current-activity/2021/07/04/cisa-fbi-guidance-msps-and-their-customers-affected-kaseya-vsa

Cybersecurity and Infrastructure Security Agency, 2023. Critical Infrastructure Sectors. *Critical Infrastructure Security and Resilience*. Available from: www.cisa.gov/topics/critical-infrastructure-security-and-resilience/critical-infrastructure-sectors

Dailey, N., 2021. The Hackers that Attacked a Major US Oil Pipeline Say It Was Only for the Money – Here's What to Know About DarkSide. *Business Insider*, 10 May. Available from: www.businessinsider.com/pipeline-cyber-attack-darkside-hacker-group-shutdown-ransomware-money-politics-oil-2021-5

DARK Reading, 2021. 60% of Organizations Would Consider Paying in the Event of a Ransomware Attack. *DARK Reading*, 18 June. Available from: www.darkreading.com/attacks-breaches/60-of-organizations-would-consider-paying-in-the-event-of-a-ransomware-attack

Deloitte Center for Financial Services, 2020. Reshaping the Cybersecurity Landscape. *Deloitte Insights*. Available from: www2.deloitte.com/content/dam/insights/us/articles/6507_Cybersecurity-FS-ISAC/DI_2020-FS-ISAC-Cybersecurity.pdf

Dutton, J., 2021. Beef Shortage Update: Prices Rise As Plants Recover From JBS Cyberattack. *Newsweek*, 3 June. Available from: www.newsweek.com/beef-prices-rise-plants-recover-jbs-hack-1597153

Federal Bureau of Investigation, 2017. 2016 Internet Crime Report. *IC3*. Available from: www.ic3.gov/Media/PDF/AnnualReport/2016_IC3Report.pdf

Federal Bureau of Investigation, 2019. IC3 Annual Report Released: Report Shows Cyber-Enabled Crimes and Costs Rose in 2018. *FBI News*, 22 April. Available from: www.fbi.gov/news/stories/ic3-releases-2018-internet-crime-report-042219

Federal Bureau of Investigation, n.d. Ransomware. *Scams and Safety*. www.fbi.gov/scams-and-safety/common-scams-and-crimes/ransomware

Gatlan, S., 2021. US Unites 30 Countries to Disrupt Global Ransomware Attack. *Bleeping Computer*, 1 October. Available from: www.bleepingcomputer.com/news/security/us-unites-30-countries-to-disrupt-global-ransomware-attacks/

Glover, C. 2022. The Ransomware Crisis is Making Cyber Insurance Harder to Buy. *Tech Monitor*, 26 January. Available from: https://techmonitor.ai/technology/cybersecurity/cyber-insurance

Government Accountability Office, 2021. Cyber Insurance: Insurers and Policyholders Face Challenges in an Evolving Market. *Report to Congressional Committees,* May. Available from: www.gao.gov/assets/gao-21-477.pdf

House Armed Services Committee, 2022. Final Text Summary of the National Defense Authorization Act for Fiscal Year 2022. *U.S. House of Representatives.* Available from: https://rules.house.gov/sites/democrats.rules.house.gov/files/17S1605-RCP117-21-summary.pdf

Hudak, T., 2021. The True Cost of a Ransomware Attack. *DARKReading*, 3 June. Available from: https://beta.darkreading.com/vulnerabilities-threats/the-true-cost-of-a-ransomware-attack

Intel 471, 2021. The Moral Underground? Ransomware Operators Retreat After Colonial Pipeline Hack. *Intel471*, 14 May. Available from: https://intel471.com/blog/darkside-ransomware-shut-down-revil-avaddon-cybercrime

i-Scoop, n.d. Operational Technology (OT) – Definitions and Differences with IT. *i-Scoop.* Available from: www.i-scoop.eu/industry-4-0/operational-technology-ot/

Kochovski, A., 2022. Ransomware Statistics, Trends and Facts for 2022 and Beyond. *Cloudwards*, 3 January. Available from: www.cloudwards.net/ransomware-statistics/

Krauss, C. and Sanger, D. E., 2021. Gasoline Buying Fever Rages as Pipeline Company Begins Restart. *The New York Times*, 12 May. Available from: www.nytimes.com/2021/05/12/business/energy-environment/pipeline-shutdown-latest-news.html

Krebs, B., 2021. DarkSide Ransomware Gang Quits After Servers, Bitcoin Stash Seized. *KrebsonSecurity*, 14 May. Available from: https://krebsonsecurity.com/2021/05/darkside-ransomware-gang-quits-after-servers-bitcoin-stash-seized/

Lepore, J., 2021. The Next Cyberattack is Underway. *The New Yorker*, 1 February. Available from: www.newyorker.com/magazine/2021/02/08/the-next-cyberattack-is-already-under-way

Midler, M, 2020. Ransomware as a Service (RaaS) Threats. *SEI Blog*, 5 October. Available from: https://insights.sei.cmu.edu/blog/ransomware-as-a-service-raas-threats/

Myre, G., 2021. FBI Says Darkside Ransomware Is Responsible for Attack on U.S. Pipeline. *National Public Radio*, 11 May. Available from: www.npr.org/2021/05/11/995751021/fbi-says-darkside-ransomware-is-reponsible-for-attack-on-u-s-pipeline

Nadeau, M., 2020. General Data Protection Regulation (GDPR): What You Need to Know to Stay Compliant. *CSO*, 12 June. Available from: www.csoonline.com/article/3202771/general-data-protection-regulation-gdpr-requirements-deadlines-and-facts.html

Patel, S., 2021. DHS Issues Pipeline Cybersecurity Directive but Industry Championing FERC Mandatory Standards. *Power*, 31 May. Available from: www.powermag.com/dhs-issues-pipeline-cybersecurity-directive-but-industry-championing-ferc-mandatory-standards/

Perlroth, N. and Satariano, A., 2021. Irish Hospitals Are Latest to Be Hit by Ransomware Attacks. *The New York Times*, 20 May. Available from: www.nytimes.com/2021/05/20/technology/ransomware-attack-ireland-hospitals.html

Perlroth, N., et al., 2021. Pipeline Investigation Upends Idea that Bitcoin Is Untraceable. *The New York Times*, 9 June. Available from: www.nytimes.com/2021/06/09/technology/bitcoin-untraceable-pipeline-ransomware.html

Platsis, G., 2022. What CISOs Should Know About CIRCA Incident Reporting. *Security Intelligence*, 8 December. Available from: https://securityintelligence.com/articles/what-cisos-should-know-circia-incident-reporting/

Pomerleau, P. and Maimon, D., 2020. *Evidence-Based Cybersecurity: Foundations, Research, and Practice*. Boca Raton, FL: CRC Press.

Popovici, M., 2023. Food Giant Dole, Victim of Ransomware Attack. *Heimdal Security*, 24 March. Available from: https://heimdalsecurity.com/blog/food-giant-dole-victim-of-a-ransomware-attack/

Powell, B., 2005. Is Cybersecurity a Public Good? Evidence from the Financial Services Industry. *Journal of Law, Economics, and Policy*, 1(2), 487–510.

Reed, J., 2022. Costa Rica State of Emergency Declared After Ransomware Attacks. *Security Intelligence*, 16 November. Available from: https://securityintelligence.com/news/costa-rica-state-emergency-ransomware/

Reuters, 2021. White House Contact Russia After Hack of World's Largest Meatpacking Company. *The Guardian*, 1 June. Available from: www.theguardian.com/technology/2021/jun/01/jbs-meatpacking-ransomware-hack-russia-white-house

Ridge, T., 2003. Remarks by Secretary of Homeland Security Tom Ridge at the 2003 National Cyber Security Summit. *Department of Homeland Security*, 3 December. Available from: www.techlawjournal.com/security/20031203.asp

Robbins, R., 2021. Meat Processor JBS Paid $11 Million in Ransom to Hackers. *The New York Times*, 9 June. Available from: www.nytimes.com/2021/06/09/business/jbs-cyberattack-ransom.html

Rosenbaum, E., 2021. JBS Cyberattack: From Gas to Meat, Hackers Are Hitting the Nation, and Consumers, Where It Hurts. *CNBC*, 2 June. Available from: www.cnbc.com/2021/06/02/from-gas-to-burgers-hackers-hit-consumers-where-it-hurts.html

Sanger, D. E. and Perlroth, N., 2021. The FBI Identified DarkSide as a Colonial Pipeline Hacker. *The New York Times*, 10 May. Available from: www.nytimes.com/2021/05/10/us/politics/pipeline-hack-darkside.html

Schwirtz, M. and Perlroth, N., 2021. DarkSide, Blamed for Gas Pipeline Attack, Says It Is Shutting Down. *The New York Times,* 14 May. Available from: www.nytimes.com/2021/05/14/business/darkside-pipeline-hack.html

Simpson, A., 2021. BCPS Ransomware Recover Efforts Come With $8.1 Million Price Tag. *Fox 5 News Baltimore*, 15 June. Available from: https://foxbaltimore.com/news/local/bcps-ransomware-recovery-efforts-come-with-81-million-price-tag

Smith M. and Monken, J., 2021. The Colonial Pipeline Hack Shows We Need a Better Federal Cybersecurity Ecosystem. *The Modern War Institute at West Point*, 21 May. Available from: https://mwi.usma.edu/the-colonial-pipeline-hack-shows-we-need-a-better-federal-cybersecurity-ecosystem/

U.S. Department of Defense, 2020. Fiscal Year 2019 Operational Energy Annual Report. *Office of the Undersecretary of Defense for Acquisition and Sustainment*, 1 March. Available from: www.acq.osd.mil/eie/Downloads/OE/FY19%20OE%20Annual%20Report.pdf

11 Unconventional Warfare in the Information Environment

Otto C. Fiala and Jim Worrall

To win one hundred victories in one hundred battles is not the acme of skill. To subdue the enemy without fighting is the acme of skill.

(Sun Tzu; McNeilly 2001, p. 18)

Introduction

The term unconventional warfare (UW) is commonly misunderstood and subject to frequent calls for revision (Livermore 2017). The United States Army's Special Operations Command has historically had purview over this activity. In 2016, it defined UW as s "activities conducted to enable a resistance movement or insurgency to coerce, disrupt, or overthrow a government or occupying power by operating through or with an underground, auxiliary, and guerrilla force in a denied area" (UW Pocket Guide 2016). At the Joint level (involving all the armed services) today, it remains part of special operations and continues to be similarly described (JP 1-02 2017). Both historically and doctrinally, UW has always had three objectives: to coerce, disrupt, or overthrow an adversary. However, too often, the focus of UW discussions is centered on support to indigenous forces to enable the overthrow of an oppressive regime – rarely is the focus on its use for the purposes of coercion or disruption. This is despite the incredible difficulty and improbability of actually overthrowing a government – let alone the international law problem. Yet, an improved understanding of UW and, more specifically, the operational or strategic disruption conducted either by or against an adversary will allow us to better recognize when our adversaries (before their activities graduate them to "enemies") apply historic UW techniques and intent through the most sophisticated methods of the early 21st century allowing engagement with whole societies, against the United States or our allies and partners.

This chapter examines UW from the perspective of great power competition and the current state of persistent engagement with our two near-peer competitors, Russia and China. Today's perpetual competition manifests great power competition as defined in the 2017 National Security Strategy (Trump 2017) and revisited in the Interim National Security Strategic Guidance released in 2021 (Biden 2021).

DOI: 10.4324/9781003425304-12

Specifically, this chapter investigates Russian and Chinese integrated operations in the information environment (OIE) and cyber domain as both a form of continuous competition that falls below the threshold of open armed conflict and also as shaping operations. We assess Russia's and China's OIE effects and capabilities, our understanding of their OIE, and responses to their aggression, and make recommendations.

The Problem

Our adversaries' OIE are executed with specific purposes. However, the United States has failed to qualify these activities properly and the threat they pose to national security, affecting our ability to counter and deter the threat adequately. Much of the problem is rooted in the terminology applied toward understanding the problem of Russian and Chinese OIE. Terms like "malign activities," "influence operations," and "disinformation" have all been used to describe adversarial OIE, but none of the terms adequately capture OIE's *disruptive* effects or the intent of these adversaries. The inadequacy of current terminology and the ways that certain terms can confuse rather than clarify their OIE is illustrated by the examples of Eastern Europe and Taiwan.

Russian OIE in Eastern Europe and the Baltics

Over the past decade, Russia has worked (Votel et al. 2016) to expand its sphere of influence and control into former Soviet or Warsaw Pact territory to the greatest extent possible without triggering a NATO Article 5 response (NATO 1949). For almost the past decade and prior to its invasion of February 2022, Russia took calculated measures to prevent Ukraine from controlling its easternmost territory (Crisis Group Europe and Central Asia 2016) through its continued involvement in the ongoing conflict (BBC 2014) over Crimea (Simpson 2014) and the Donbas. Russian activity in the region effectively derailed and ultimately prevented Ukraine's bid to join NATO and the European Union.

Most recently, as part of its latest conflict inflicted upon Ukraine, on 24 February 2022, Putin used the full force of his military in an attempt to gain control over Ukraine and re-unite Russians and Ukrainians as "one people," and launched his "special military operation" (invasion) (Dickinson 2022). On the same day as the ground invasion, Russia attacked Viasat terminals across Ukraine to disrupt internet connectivity and, thus communication as it opened its ground attack. The cyber-attack launched destructive "wiper" malware called Acid Rain against Viasat modems and routers, erasing all the data on those systems. When the machines were rebooted, they were permanently disabled. This effectively destroyed thousands of terminals, with effects reaching the European Union by interfering with internet-connected wind farms in central Europe (O'Neill 2022).

Long before the invasion, Russia also prepared human agents to influence the information environment. A particularly egregious example was Andriy Derkach, a People's Deputy in the Ukrainian Parliament with a long history of working with the Russians and advancing their interests. As part of an effort to undermine relations between the United States and Ukraine, in 2019–2020, he made public several documents (likely forged) and audio recordings of President Poroshenko's conversations with former U.S. Vice President Biden and President Putin, suggesting systematic U.S. interference in Ukrainian politics and implying corrupt activities by high-ranking U.S. officials in Ukraine (U.S. Department of Treasury 2021). His activities were obviously meant to disrupt relations between the United States and Ukraine and in hindsight can be seen as part of a larger campaign of disruption. This attempt at disruption through an operation in the information environment was intended to gain political and psychological maneuver room for Russia.

Additionally, Russia's threat to the Baltic states (Chang 2017) is increasing in ways that do not involve the use of military force (Episkopos 2021). Instead, Russia has significantly increased its ability to access and influence the information environment. An important aspect of operating in the information environment is the ability to conduct operations in and through cyberspace. To this end, Russia has invested heavily in cyber capabilities, increasing its ability to conduct destructive attacks in cyberspace and exert pressure on Eastern Europe and the Baltic states. For example, in 2006, sparked by the Estonian decision to remove a Russian World War II memorial, the "Bronze Soldier" (Cavegn 2017), Russia conducted an intensive, three-week-long cyberattack against Estonia (McGuinness 2017). Later, in 2008, Russia conducted a coordinated military offensive that included cyber and kinetic attacks against Georgia (Shakarian 2011) to seize Abkhazia and North Ossetia. And in June 2017, the Russians conducted the most destructive (Banerjea 2018) cyberattack to date, using malware known as *NotPetya* against Ukraine's critical infrastructure and other targets.

The relative effectiveness of those attacks lies in great contrast to the Russian conflict in Ukraine after its 2022 invasion. During this conflict, private entities such as Microsoft and Starlink have provided secure cloud storage services, cybersecurity, and internet connectivity. Direct attacks against these NATO-based commercial entities by Russia would threaten to expand the war, likely unfavorably to Russia.

Eastern European and Baltic Resilience in the Information Environment

Due to the ongoing threat of Russian aggression, the Baltic states and Sweden distribute detailed guidance (Swedish Civil Contingencies Agency 2022) to their populations regarding what to do in the event of a Russian military invasion. However, the Baltic nations are similarly well prepared for cyber engagements. In 2006, before the "Bronze Soldier" attack discussed above, Estonia presciently created its first Cyber Emergency Response Team, designed to mitigate the effects of Russian denial-of-service attacks. After the attack, Estonia further expanded its cyber incident response capabilities and created a cyber defense unit within the Estonian Defense League. Not surprisingly, Estonia continues to exhibit a high and expanding degree of cyber resiliency and is internationally known for the NATO Cooperative Cyber Defence Center of Excellence established in 2008. Latvia soon followed Estonia's lead and created NATO's Strategic Communications Center of Excellence.

Another, distinctly Baltic, development in cyber defense occurred in Lithuania: in 2015, the Lithuanians created the "elves" (Kojala 2020) a volunteer network of citizens charged with monitoring the internet for Russian disinformation. The Czech Republic is also an innovative leader in combating Russian disinformation and created the Centre Against Terrorism and Hybrid Threats in 2017. Prior to its accession to NATO in April 2023, Finland similarly created the European Centre of Excellence for Countering Hybrid Threats. This, coupled with support from U.S. Cyber Command conducting foreign internal defense activities and assisted by the U.S. Department of Homeland Security's (DHS) Cybersecurity and Infrastructure Security Agency (CISA), makes the Baltic and Eastern European nations very competent and capable defenders in cyberspace and much of the information environment. This gradual building of cyber defenses likely ensured that Russia's second major cyber-attack on Estonia, in August of 2022, executed by the Russian hacker group Killnet, in response to the Estonian government moving Soviet monuments out of the heavily ethnic Russian-populated city of Narva, was barely noticed by the average Estonian (Sytas 2022).

Russian OIE in the United States

Russia has also extended its cyber reach into the domestic United States. On 16 March 2021, the Office of the Director of National Intelligence released the declassified intelligence community assessment (Intelligence Community Assessment 2021) of foreign threats to the 2020 U.S. federal elections. The report specifically identifies Russia as interfering in those elections with the intent of exacerbating preexisting social divisions and undermining American's trust and confidence in democratic institutions. The intelligence community assessment states that a range of Russian government organizations participated in the influence operations, using Russian state and proxy actors to affect public perceptions in a strategic and calculated manner (Vartanova 2021). A key element of Moscow's strategy was its use of proxies linked to Russian intelligence that pushed influence narratives – including misleading or unsubstantiated allegations against political figures, media organizations, U.S. officials, and even private U.S. citizens.

Russia's influence campaigns during the 2020 election cycle are only a small part of Russian OIE, which includes the hacking of the Democratic National Committee in 2016 (Lewis 2016), the recent SolarWinds hack (Palmer 2021), and other activities. Russia's ability to reach directly into the U.S. domestic information environment enables Russian shaping operations and gives the Russians access to, and influence in, the U.S. public sphere.

Russian OIE and Disruption

Broadening the aperture from the tactical application of cyber warfare to the more expansive and inclusive zone of OIE allows us to better understand the characteristics of Russian aggression in the information space. Understanding OIE is crucial to exerting dominance in the information environment. For example, the inability to comprehend the strategic scope of Russian OIE in Crimea in 2014 (United States Army Special Operations Command, no date) prevented a quick response from Ukraine and Western allies and partners, which allowed Russia enough freedom of movement to shape the information environment before quickly mobilizing its ground forces – a move that took the Ukrainians and the West by surprise.

In short, Russian OIE is ever present as either a main or a supporting effort. OIE with the intent of disruption was the main effort for Russian interference in the 2016 and 2020 U.S. presidential elections. In Georgia, Russian OIE was a precursor to armed conflict and was used to shape the information environment before ground troops were mobilized. History dictates that Russia will continue to make use of OIE in future engagements – as a supporting or a main effort – including in any action Russia chooses to take against NATO.

Chinese OIE against Taiwan

Over the past decade, the People's Republic of China (PRC) has gradually but inexorably overtaken Russia as America's most challenging global competitor. Chinese learning from Russia, added to its own initiatives, extends this into OIE. The PRC has had the advantage of watching Russian OIE married with kinetic activity, particularly in Crimea in 2014 and its more expansive conflict in Ukraine beginning in February 2022, and its effect on and reactions from the West. Additionally, the PRC has directly engaged in OIE concerning the outbreak of COVID-19 and its likely association with the Wuhan Institute of Virology (Office of the Director of National Intelligence, no date). It also engaged in OIE in Hong Kong (Bureau of East Asian and Pacific Affairs 2023) and the loss of freedom there in 2020 under the imposition of a new National Security Law, during a time when most of the West was more concerned with defending itself against a new virus. These events, from which it has learned, directly build into its ability to target and conduct OIE against Taiwan in an ever-increasing fashion as it moves closer to the day it coercively demands re-unification, under the PRC.

People's Liberation Army Strategic Support Force

In 2015, China's People's Liberation Army (PLA) initiated many structural reforms. In particular, the PLA created a Strategic Support Force (SSF), centralizing most PLA space, cyber, electronic, and psychological warfare capabilities, and reporting to the Central Military Commission (Costello and McReynolds 2018). Yet, the SSF has not simply provided a unifying command function for complementary or overlapping activities (Costello and McReynolds 2018). It also seems to have incorporated the PLA's psychological and political warfare missions. This includes encapsulation of a significant portion of the Three Warfares concept developed in the early 2000s (Livermore 2018). They include "public opinion warfare" to influence perceptions, "psychological warfare" relying on intimidation, and "legal warfare" to battle in the courts. These "Three Warfares" can operate separately, prior to, or in tandem with military operations. For example, China could spread disinformation, while sailing naval vessels in Taiwan's contiguous zone and conduct military exercises, intending to cause mass panic. The SSF is responsible for conducting such influence operations, as well as conducting cyber and electronic warfare, and space operations, giving them the combined warfare tools to successfully operationalize these efforts under SSF control. The SSF appears to be deeply engaged in operations in the psychological and cognitive domains (Nemoto et al. 2022). The Three Warfares concept is intended to control perceptions and shape narratives to advance Chinese interests while undermining the interests of opponents (Kania 2016). The space, cyber, and electromagnetic spectrum domains are the primary conduits through which they conduct their operations. If an adversary military force, using informationized system-of-systems infrastructure can be denied the use of these domains, then that adversary force cannot properly function (Costello and McReynolds 2018). Thus, the SSF is responsible for achieving information dominance during conflict.

PRC Attacks against Taiwan

The PRC has long made it known that it seeks to bring Taiwan under its control, viewing it as a renegade entity that should be rightfully an uncontested part of the PRC. In so doing, the PRC would rather win this objective without fighting. Yet, whether this can be achieved without fighting is not for evaluation here. The PRC has been engaging in aggressive information and psychological warfare against the citizens of Taiwan while laying the groundwork in the IE around the world for many years and has been gradually increasing this activity. Its attacks in the IE rise to the level of cognitive warfare by using information, misinformation, disinformation, or a combination of them to achieve societal influence in the cognitive domain (Reding and Wells 2022). This form of warfare is in keeping with the Chinese theory of winning without fighting. Yet, such activity typically requires a very lengthy horizon for effect but can be used to shape the environment in case of a decision to engage in kinetic activity. It can be used over a term of years to damage or lessen the resilience of a target society, thus making the target society more vulnerable to successful kinetic activities.

PRC attacks seem to be more sophisticated than those conducted by Russia. In 2021, during the COVID pandemic, the vaccination rate in Taiwan was low as was its vaccine stock due to the practical nonexistence of the virus on the island until an outbreak in May of that year (Gan and Cheung 2021). When Taiwan then sought to acquire vaccines from abroad, the PRC accused Taiwan of seeking independence and vigorously fought Taiwan's ability to acquire vaccines from other countries to limit its vaccine supply and to shape an image of the Taiwanese government as incompetent and unreliable (Zhong and Schuetze 2021). In December 2020, Taipei arranged to purchase five million vaccine doses from Pfizer-BioNTech (BNT), but BNT suddenly canceled without justification. When Taiwan's Minister of Health and Welfare revealed this publicly, BNT stated that the vaccine deal remained unchanged. It was later confirmed that the deal had been originally breached by BNT due to PRC intervention to prevent Taiwanese purchase of the vaccine. In a continuation of this cognitive warfare, China then publicly stated its willingness to provide vaccines to Taiwan. However, the PRC worked to block deliveries to make the Taiwanese government appear incompetent and cause distrust of it among the Taiwanese people. China then targeted a Taiwanese-developed vaccine by exaggerating its side effects to increase public skepticism in the vaccine and discourage vaccination, leading to further virus-related complications for Taiwan. The PRC also fabricated accusations of senior Taiwanese officials profiteering from the situation. This was combined with sending other fake news into Taiwan to divide its population in an attempt to demoralize it and undermine its confidence in the independence-minded Democratic Progressive Party (DPP) government (Tsung-Chi Yu and Ho 2022).

During 2022, the government of Taiwan experienced approximately 30 million cyberattacks a month, mostly attributable to China (Kagubare 2023). According to the National Institute for Defense Studies (NIDS), a Japanese Defense Ministry think tank, between September 2019 and August 2020, Taiwan sustained 1.4 billion cyberattacks, targeting political, economic, and military entities, in apparent attempts to destroy or steal data. The Chinese also attempt to spread misinformation and disinformation under the guise of Taiwanese users. In April 2020, during the height of the pandemic, Tedros Adhanom Ghebreyesus, the director-general of the World Health Organization, said he had been subject to racial slurs on the internet from Taiwanese accounts. Taiwan responded by saying the slurs originated from China but were made under the pretense of being Taiwanese people as part of an influence operation to control the thought process of Director-General Tedros (Nemoto et al. 2022).

Taiwanese Resilience against PRC Attacks

As noted above, Taiwan must defend itself against approximately thirty million cyberattacks per month, with at least half of them attributable to China (Hille 2019). In a recent effort to defend against such attacks, Taiwan President Tsai Ing-wen unveiled a new National Institute of Cyber Security (NICS) on 10 February 2023. The institute falls under the Ministry of Digital Affairs, which is a new cabinet-level ministry formed in August 2022. The institute will engage in research and make policy recommendations regarding cyber and data security, assist both private and public entities during major cybersecurity incidents, and assist with the implementation of a secure government data sharing system known as the "T-Road" system. Further, NICS is integrated into the Taiwan Computer Emergency Response Team/Coordination Center (TWCERT/CC), which is the government-led incident response center for major private sector data breaches (Gorin 2023). It will also soon establish the Taiwan Academic Cybersecurity Center to be operated by the National Science and Technology Council (Li-hua 2023).

Taiwan's cybersecurity initiatives for critical infrastructures and building digital resilience also include the promotion of international standards for semiconductor supply chain information security (SEMI E187), the implementation of a zero-trust framework, app cybersecurity testing, cybersecurity standards for chips and the Internet of Things, and the establishment of the Shalun Information Security Service Base (Ocampo 2023). Specifically, zero trust is an approach where access to data, networks, and infrastructure is kept to what is minimally required, and the legitimacy of that access must be continuously verified (CISA 2023a,b), while Shalun is a small "science city" established in Tainan focused on cybersecurity, smart technology, energy, and the development of new technological industries (Chiang 2021).

Taiwan also issued the Cyber Security Management Act (MoJ 2018), using it to build an inter-ministerial information security joint defense framework. The country has made progress in safeguarding key infrastructure, performing field audits to ensure the efficiency of protection measures, and conducting the biennial Cyber Offence and Defence Exercise (CODE) since 2013 (Ocampo 2023; Hille 2019). It is unclear whether a lack of a major attack, such as those suffered by Estonia from Russia, is a result of effective Taiwanese defenses or Chinese capability or decisions.

Chinese OIE in the U.S.

At the U.S. Cybersecurity and Infrastructure Security Agency website (CISA 2023a,b), the agency cites the 2023 annual threat assessment by the Office of the Director of National Intelligence regarding the cyber threat posed by the PRC:

> China probably currently represents the broadest, most active, and persistent cyber espionage threat to U.S. Government and private sector networks. China's cyber pursuits and its industry's export of related technologies increase the threats of aggressive cyber operations against the US homeland … China almost certainly is capable of launching cyberattacks that could disrupt critical infrastructure services within the United States, including against oil and gas pipelines, and rail systems.

In May of 2023, Chinese hackers used "stealthy" malware to attack critical infrastructure on American military bases on the island of Guam in one of the largest known cyber espionage campaigns against the United States (Ritchie 2023). Microsoft revealed that it tracked the attacks to a group of what it believes to be Chinese state-sponsored hackers who have, since 2021, carried out a broad hacking campaign. Microsoft has named the group Volt Typhoon. That group has targeted critical infrastructure systems in the United States and Guam, including communications, manufacturing, utilities, construction, and transportation. The hacking did not appear to be disruptive, but Microsoft warned that the nature of the group's targeting may yet enable future disruption because the hackers are likely developing capabilities that could disrupt critical communications infrastructure between the United States and Asia (Greenberg and Newman 2023).

Google-owned cybersecurity firm Mandiant has also tracked Volt Typhoon's intrusions and offered similar warnings regarding their intent, especially since the hacking lacked a clear connection to intellectual property or policy information typical of an espionage operation. Mandiant's conclusion is similar to that of Microsoft and warns of the group's focus on critical infrastructure, which could lay the groundwork for a disruptive or destructive attack since gaining the access necessary for a disruptive attack usually requires months of advanced work (Greenberg and Newman 2023).

Distinguishing between espionage, cyberattack preparation, and imminent cyberattack is difficult to discern with Chinese hackers given their limited disruptive events, making the PRC's capabilities for disruptive and destructive attacks difficult to evaluate. In a conflict with the PRC, U.S. military facilities on Guam, including ports and air bases would be crucial to any Western response.

In these activities directly targeting U.S. facilities, the PRC is targeting critical infrastructure used for both civilian and military purposes, thus eroding the principles of the law of war. It is necessary to also understand that under international law, an attack against the U.S. territory of Guam is the equivalent of an attack on any state of the United States. The law of armed conflict ordinarily forbids targeting civilian objects, such as civilian property and infrastructure. However, many computer networks are "dual use" used for both civilian and military purposes. During war they may be targetable based on their nature, purpose, and use. However, combatants must still comply with the principles of the law of war, such as military necessity, proportionality, and avoiding unnecessary suffering. The PRC's escalatory activities against both Taiwan and the United States challenge the concepts of peacetime and wartime and bring about questions regarding the law of armed conflict (Goldenziel 2023). China may be conducting significant cyber activities as part of shaping operations in preparation for conflict while also allowing for the possibility that the shaping efforts themselves, particularly against Taiwan, may themselves bring about the intended objective of winning without fighting. There is perhaps an appropriate adage for this situation; to defend yourself, you must first realize that you are being attacked.

PRC Conceptions of OIE

China has built upon both the United States and Russian understanding of OIE through its drive for Intelligentized Warfare (Jacobs and Carley 2022). In particular, by combining cyber data with cyber, electromagnetic and space warfare, its objective is to steer its adversary's perception in a desired direction in order for that adversary to choose a course of action or objective that China desires. This is a step beyond the older Russian concept of reflexive control (Sprang 2018). Intelligentized Warfare integrates more traditional information warfare with big data, social networks, media, and commerce. China is moving beyond its "system-of-systems" framework, by seeking algorithmic advantage to achieve an operational advantage (Kania 2019b). China is beginning to marry artificial intelligence with emerging technologies to gain an advantageous position in what it sees as the next Revolution in Military Affairs (RMA) (Kania 2019a). China's New Generation Artificial Intelligence Development Plan incorporated a "Whole-of-Nation" approach to create and harvest technical talent and invest in disruptive technologies. Aspects of this nascent domain include developing "human-out-of-the-loop" automation, increased 5G connectivity and data as the foundation for this new intelligentized battlefield (Jacobs and Carley 2022; Kania 2019b).

U.S. Cognitive Tools

Comparatively, Russian and Chinese OIE far exceed anything similarly conducted by the United States. The Irregular Warfare Annex to the 2018 National Defense Strategy affirmed that the United States is engaged in global competition to advance its interests and gain enduring strategic advantage (United States Department of Defense 2020). Additionally, the Department of Defense's (DoD) approach to irregular warfare encompasses both competition and conflict with potential state adversaries. The annex encourages DoD to take the initiative and engage in asymmetric and indirect efforts to erode an adversary's power, influence, and will. DoD should also adopt proactive, dynamic, and unorthodox approaches to irregular warfare to shape, prevent, and prevail against adversaries. In doing so, it will maintain favorable regional balances of power. Activities like information support operations, cyberspace operations, countering threat networks, counter–threat finance, civil–military operations, and security cooperation can help shape the information environment. Finally, acknowledging that state adversaries and their proxies increasingly seek to prevail in the information environment using irregular warfare tactics, the annex indicates that irregular warfare tactics can also proactively shape conditions to the United States' advantage in great power competition.

Even though UW – as currently defined – requires underground, auxiliary, and guerrilla forces, a key UW objective is disruption, and we argue that disruption is the objective of most Russian and Chinese OIE. Additionally, in today's interconnected world – where human perception is easily influenced – their use of witting and unwitting proxies (instead of underground, auxiliary, and guerrilla forces) to conduct OIE achieves a similar effect. Indeed, though its efforts could not affect the outcome, societal disruption per UW is what Russian OIE accomplished during the 2020 U.S. elections.

Russian OIE against the United States has included influence operations to undermine public confidence in the U.S. electoral process and to exacerbate current social and political divisions within the U.S. population. Russia and China also use proxy actors to affect public perceptions and proxies linked to intelligence organizations push influence narratives through social media. Because cognitive perceptions are primarily formed and shaped via interconnected cyber and information environments, Russia and China do not need the traditional components of UW (underground, auxiliary, and guerrilla forces) to disrupt.

Yet, the terms "disruption" and "disrupt" are not defined by the DoD Dictionary or the Irregular Warfare Annex despite being a UW objective targeted at a government or occupying power. The U.S. Army does offer a tactical-level definition of "disrupt," but that definition fails when the intended effect is political. Joint Doctrine Note (JDN) 1-19 of 2019 (Joint Chiefs of Staff 2019), though not an authoritative document, contains a suggested definition for disrupt: "temporarily interrupt the enemy's activities or the effectiveness of enemy organizations by interdiction, subversion, or coercion." In this case, JDN 1-19's use of the word "enemy" (instead of "adversary") is intentional, as the definition is intended to apply to armed conflict, not activities or competition executed below the threshold of war.

Further, the DoD Dictionary (United States Department of Defense 2017) defines subversion as "actions designed to undermine the military, economic, psychological, or political strength or morale of a governing authority." This definition includes a political effect and helps explain the intent behind Russian and Chinese OIE for many of the historical examples provided above. Ultimately, their OIE has been aimed at undermining political strength or morale with the further intent of causing political disruption.

Section 1097 of the 2017 National Defense Authorization Act also acknowledges the political element of UW, mandating that DoD develop a strategy "to counter unconventional warfare threats posed by adversarial state and non-state actors" (NDAA 2017). However, it remains difficult to develop a strategy to counter an activity when doctrinal definitions fail to capture the true intent of an adversary's actions.

During the Cold War, the United States and the Soviet Union selectively applied force to prevent further escalation, maintaining a constant state of tension below the level of armed conflict. In fact, many (Votel et al. 2016) place the Cold War and our current great power competition squarely in a segment of the conflict continuum known as the "gray zone" (Barno and Bensahel 2015). The gray zone is characterized by "intense political, economic, informational, and military competition more fervent in nature than normal steady-state diplomacy, yet short of conventional war" (Votel et al. 2016). But the concept is hardly new. George Kennan identified U.S. Cold War foreign policy as "political warfare" (Kennan 1948), which was recently reimagined as "unconventional warfare in the gray zone" (Votel et al. 2016). In the gray zone, UW is a shaping effort or a main effort, and UW tactics are often used prior to the deployment of conventional forces.

Recommendations

Though disruption as operational intent may sound innocuous, it is not. As a nation, our inability to recognize adversarial OIE as a form of UW intended to undermine the United States, allied and partner nation democratic institutions, and sow civil discord places us at a disadvantage when countering or executing OIE ourselves. During the Cold War, we engaged in a nuclear arms race and the intent was deterrence. The cost of war became too great, and the doctrine of mutually assured destruction prevented escalation. Presently, despite being in an era that occupies the same segment on the conflict continuum as the Cold War, Russian, and Chinese OIE continue undeterred. Both Russia and China have strategically developed and deployed capabilities that directly engage populations, and by extension, their representative governments. Their OIE actively disrupts the flow of information, disrupts democratic processes and institutions, and undermines preexisting social norms.

To prevent disruption operations from Russian and Chinese OIE, we recommend several steps the United States should take to deter future OIE and to conduct OIE ourselves, if necessary. Like the mutually assured destruction doctrine of the Cold War, the United States must arm itself with weapons designed to counter and deter adversary OIE capabilities. Where the Cold War required man-portable anti-tank weapons to counter Soviet armored forces' numerical advantage, the current gray zone competition requires a similar doctrine for the information environment to curb the threat of escalation and keep engagements below the threshold of armed conflict.

DoD should revise professional military education to include UW and foster widespread understanding of UW as a form of warfare marked by persistent engagement and requiring constant attention. The competition continuum, as outlined in JDN 1-19, should be revised by moving "disrupt" as an activity of armed conflict to an activity that also takes place in competition below the threshold of armed conflict. Further, to emphasize the impact of adversarial OIE and its disruptive effect on our democracy, the revision of "disrupt" should be issued within authoritative guidance, concurrent with a revision of joint doctrine. Within Army doctrine specifically, an adjustment to the definition of "insurgency" is required and should include nonviolent tactics, techniques, and procedures (e.g., organized use of subversion and violence or nonviolence), to account for potential kinetic effects from OIE.

Moreover, new policies are required to emphasize the doctrinal changes. New policy should recognize adversarial disruption through OIE as UW (subversion to disrupt), opening the door for the United States to conduct similar offensive operations in response. To maintain flexibility, any new policy should authorize OIE for disruption by, with, and through witting or unwitting indigenous organizations – offensively or defensively. Finally, to effectively use OIE and better counter its effects on our own population and government, the United States needs to normalize the use of OIE in the gray zone to disrupt our adversaries and dominate the information space in great power competition.

Finally, collaboration with allied and partner nations and commercial entities is also critical to successful deterrence and defense. The Russian conflict in Ukraine since February 2022 has taught us the criticality of collaboration across nations and commercial entities to defend against attacks in cyberspace and the cognitive dimension. Ukraine would have no secure cloud-based services or internet connectivity without the availability of rapid and effective commercial responses to its needs. Vital to that successful response was the fact that those commercial entities were based in strong partner nations. Significant attacks, such as kinetic attacks, against them could swiftly expand the war to the detriment of Russia. Therein also lies an important lesson for China.

References

Banerjea, A., 2018. NotPetya: How a Russian Malware Created the World's Worst Cyberattack Ever. *Business-Standard*, 27 August. Available from: www.business-stand ard.com/article/technology/notpetya-how-a-russian-malware-created-the-world-s-worst-cyberattack-ever-118082700261_1.html

Barno D., and Bensahel N., 2015. Fighting and Winning in the "Gray Zone." *War on the Rocks*, 19 May. Available from: https://warontherocks.com/2015/05/fighting-and-winn ing-in-the-gray-zone/

BBC News, 2014. Ukraine crisis: Timeline. *BBC*, 13 November. Available from: www.bbc. com/news/world-middle-east-26248275

Biden, J., 2021. *Interim National Security Strategic Guidance*. Washington, DC: The White House. Available from: www.whitehouse.gov/wp-content/uploads/2021/03/NSC-1v2.pdf

Bureau of East Asian and Pacific Affairs, 2023. Hong Kong Policy Act Report. *U.S. Department of State*, 31 March. Available from: www.state.gov/2023-hong-kong-policy-act-report/

Cavegn, D., 2017. Monument of Contention: How the Bronze Soldier was Removed. *Estonian Public Broadcasting (EER)*, 25 April. Available from: https://news.err.ee/592 070/monument-of-contention-how-the-bronze-soldier-was-removed

Chang, F., 2017. Russia's Existential Threat to NATO in the Baltics. *Foreign Policy Research Institute*, 15 June. Available from: www.fpri.org/2017/06/russias-existential-thr eat-nato-baltics/

Chiang, S., 2021. Taiwan's Cyber Security, Smart Technology R&D Building officially opens. *Taiwan News*, 24 December. Available from: www.taiwannews.com.tw/en/news/ 4387574

Costello, J., McReynolds, J., 2018. China's Strategic Support Force: A Force for a New Era. China Strategic Perspectives, No. 13. *National Defense University Press*. Available from: https://ndupress.ndu.edu/Portals/68/Documents/stratperspective/china/china-pers pectives_13.pdf

Crisis Group Europe and Central Asia, 2016. Russia and the Separatists in Eastern Ukraine. *International Crisis Group*, briefing No. 79, 5 February. Available from: www.crisisgroup. org/europe-central-asia/eastern-europe/ukraine/russia-and-separatists-eastern-ukraine

Cybersecurity and Infrastructure Security Agency, 2023a. *China Cyber Threat Overview and Advisories*. Washington, DC: Cybersecurity and Infrastructure Security Agency. Available from: www.cisa.gov/topics/cyber-threats-and-advisories/advanced-persistent-threats/china

Cybersecurity and Infrastructure Security Agency, 2023b. *CISA Releases updated Zero Trust Maturity Model*. Washington, DC: Cybersecurity and Infrastructure Security Agency. Available from: www.cisa.gov/news-events/news/cisa-releases-updated-zero-trust-matur ity-model

Dickinson, P., 2022. Putin Admits Ukraine Invasion is an Imperial War to "Return" Russian Land. *Atlantic Council*, 22 June. Available from: www.atlanticcouncil.org/blogs/ukrai nealert/putin-admits-ukraine-invasion-is-an-imperial-war-to-return-russian-land/

Episkopos, M., 2021. World War III: If Russia Invaded the Baltics NATO Couldn't Stop Them. *The National Interest*, 15 March. Available from: https://nationalinterest.org/blog/ buzz/world-war-iii-if-russia-invaded-baltics-nato-couldnt-stop-them-180303

Gan, N., and Cheung, E., 2021. Taiwan Blames China for Slowing Down Its Access to Covid-19 Vaccines. The Reality Is More Complicated. *CNN*, 21 May. Available from: www.cnn.com/2021/05/21/asia/taiwan-covid-vaccine-intl-hnk/index.html

Goldenziel, J., 2023. China Cyberattacked The US. Corporations Are on The Front Lines. *Forbes*, 29 May. Available from: www.forbes.com/sites/jillgoldenziel/2023/05/29/china-cyberattacked-the-us-corporations-are-on-the-front-lines/?sh=4de4bdb320dd

Gorin, C., 2023. Government Unveils New Institute Focused on Cybersecurity. *Radio Taiwan International*, 10 February. Available from: https://en.rti.org.tw/news/view/id/2008988

Greenberg, A., and Newman L.H., 2023. China Hacks US Critical Networks in Guam, Raising Cyberwar Fears. *Wired*, 24 May. Available from: www.wired.com/story/china-volt-typhoon-hack-us-critical-infrastructure/

Hille, K., 2019. US and Taiwan Host Security Exercise to Boost Cyber Defence. *Financial Times*, 4 November. Available from: www.ft.com/content/7d6c78cc-fec8-11e9-b7bc-f3fa4e77dd47

Intelligence Community Assessment, 2021. Foreign Threats to the 2020 US Federal Elections. *National Intelligence Council*, 10 March. Available from: www.dni.gov/files/ODNI/documents/assessments/ICA-declass-16MAR21.pdf

Jacobs, C. S., Carley, K. M., 2022. Taiwan: China's Gray Zone Doctrine in Action. *Small Wars Journal*, 11 February. Available from: https://smallwarsjournal.com/jrnl/art/taiwan-chinas-gray-zone-doctrine-action

Joint Chiefs of Staff, 2019. Competition Continuum. *U.S. Department of Defense*, 3 June. Available from: www.jcs.mil/Portals/36/Documents/Doctrine/jdn_jg/jdn1_19.pdf [Accessed 2 February 2022].

Kagubare, I., 2023. Lawmakers introduce bill to counter Chinese cyber threats against Taiwan. *The Hill*, 20 April. Available from: https://thehill.com/policy/cybersecurity/3961117-lawmakers-introduce-bill-to-counter-chinese-cyber-threats-against-taiwan/

Kania, E. B., 2016. The PLA's Latest Strategic Thinking on the Three Warfares. *Jamestown*, 22 August. Available from: https://jamestown.org/program/the-plas-latest-strategic-thinking-on-the-three-warfares/

Kania, E. B., 2019a. Chinese Military Innovation in Artificial Intelligence. *Center for a New American Security*, 7 June. Available from: www.jstor.org/stable/resrep28742

Kania, E. B., 2019b. Testimony before the U.S.-China Economic and Security Review Commission Hearing on Trade, Technology, and Military-Civil Fusion. *Center for a New American Security*, 7 June. Available from: www.jstor.org/stable/resrep28742

Kennan, G., 1948. Policy Planning Staff Memorandum. *National Security Council*, 4 May. Available from: http://academic.brooklyn.cuny.edu/history/johnson/65ciafounding3.htm

Kojala, L., 2020. Baltic Security: The Same Challenges Remain, Even During a Pandemic. *Foreign Policy Research Institute*, 28 May. Available from: www.fpri.org/article/2020/05/baltic-security-the-same-challenges-remain-even-during-a-pandemic/

Lewis, J. A., 2016. Russia and the DNC Hacks. *Center for Strategic and International Studies*, 15 August. Available from: www.csis.org/analysis/russia-and-dnc-hacks

Li-hua, C., 2023. Cybersecurity research plans released. *Tapei Times*, 8 May. Available from: www.taipeitimes.com/News/front/archives/2023/05/08/2003799357

Livermore, D., 2017. It's Time for Special Operations to Dump "Unconventional Warfare." *War on the Rocks*, 6 October. Available from: https://warontherocks.com/2017/10/its-time-for-special-operations-to-dump-unconventional-warfare/

Livermore, D., 2018. China's "Three Warfares" in Theory and Practice in the South China Sea. *Georgetown Security Studies Review*, 25 March. Available from: https://georgeto wnsecuritystudiesreview.org/2018/03/25/chinas-three-warfares-in-theory-and-pract ice-in-the-south-china-sea/

McGuinness, D., 2017. How a Cyber Attack Transformed Estonia. *BBC*, 27 April. Available from: www.bbc.com/news/39655415

McNeilly, M., 2001. *Sun Tzu and the Art of Modern Warfare*, Oxford; New York: Oxford University Press

Ministry of Digital Affairs, 2018. Cybersecurity Management Act. *Ministry of Justice, Laws and Regulations Database*, 6 June. Available from: https://law.moj.gov.tw/ENG/LawCl ass/LawAll.aspx?pcode=A0030297

National Defense Authorization. Act for Fiscal Year 2017 (H.R. 4909; NDAA 2017, Pub.L. 114–328). *114th United States Congress*. Available from: www.congress.gov/114/plaws/ publ328/PLAW-114publ328.pdf

Nemoto, R., Ryugen, H., and Nakamura, Y., 2022. China Intensifies Disinformation, Cyberattacks on Taiwan: Report. *Nikkei Asia*, 26 November. Available from: https://asia. nikkei.com/Politics/International-relations/Taiwan-tensions/China-intensifies-disinfo rmation-cyberattacks-on-Taiwan-report

North Atlantic Treaty, 1949. *Collective Defence and Article 5*. North Atlantic Treaty Organization. Available from: www.nato.int/cps/en/natohq/topics_110496.htm

Ocampo, Y., 2023. Taiwan Building Cyber Resilience to Secure Democracy. *OpenGov*, 29 June. Available from: https://opengovasia.com/taiwan-building-cyber-resilience-to-sec ure-democracy/

Office of the Director of National Intelligence (no date). *COVID-19 Origins*. Washington, DC: National Intelligence Council. Available from: www.dni.gov/files/ODNI/documents/ assessments/Declassified-Assessment-on-COVID-19-Origins.pdf

O'Neill, P. H., 2022. Russia hacked an American satellite company one hour before the Ukraine invasion. *MIT Technology Review*, 10 May. Available from: www.technologyrev iew.com/2022/05/10/1051973/russia-hack-viasat-satellite-ukraine-invasion/

Palmer, D., 2021. SolarWinds: US and UK Blame Russian Intelligence Service Hackers for Major Cyberattack. *ZDNET*, 15 April. Available from: www.zdnet.com/article/solarwi nds-us-and-uk-blame-russian-intelligence-service-hackers-for-major-cyber-attack/

Reding, D. F., and Wells, B., 2022. Cognitive Warfare: NATO, COVID-19 and the Impact of Emerging and Disruptive Technologies. *In*: R. Gill and R. Goolsby, eds. *COVID-19 Disinformation: A Multi-National, Whole of Society Perspective*. Cham: Springer, 25–45.

Ritchie, H., 2023. Microsoft: Chinese Hackers Hit Key US Bases on Guam. *BBC*, 25 May. Available from: www.bbc.com/news/world-asia-65705198

Shakarian, P., 2011. The 2008 Russian Cyber Campaign Against Georgia. *Military Review*, November–December. Available from: www.academia.edu/1110559/The_2008_Russian_ Cyber_Campaign_Against_Georgia

Simpson, J., 2014. Russia's Crimea Plan Detailed, Secret and Successful. *BBC*, 19 March. Available from: www.bbc.com/news/world-europe-26644082

Sprang, R., 2018. Russian Operational Art, New Type Warfare, and Reflexive Control. *Small Wars Journal*, 4 September. Available from: https://smallwarsjournal.com/jrnl/art/russ ian-operational-art-new-type-warfare-and-reflexive-control

Swedish Civil Contingencies Agency, 2022. If Crisis or War Comes, (Rev) December 2022. Karlstad, Sweden: Swedish Civil Contingencies Agency. Available from: https://rib.msb. se/filer/pdf/30307.pdf

Sytas, A., 2022. Estonia Says it Repelled Major Cyber Attack After Removing Soviet Monuments. *Reuters*, 18 August. Available from: www.reuters.com/world/europe/esto nia-says-it-repelled-major-cyber-attack-after-removing-soviet-monuments-2022-08-18/

Trump, D., 2017. *National Security Strategy of the United States of America*. Washington, DC: The White House, December. Available from: http://nssarchive.us/wp-content/uplo ads/2020/04/2017.pdf

Tsung-Chi Yu, M., and Ho, K., 2022. COVID and Cognitive Warfare in Taiwan. *Journal of Asian and African Studies 2023*, 58(2), 249–273. Available from: www.ncbi.nlm.nih.gov/ pmc/articles/PMC9692172/

United States Army Special Operations Command, 2016. *Unconventional Warfare, Pocket Guide*. Fort Bragg: United States Army Special Operations Command, Deputy Chief of Staff. Available from: www.soc.mil/ARIS/books/pdf/Unconventional%20Warfare%20 Pocket%20Guide_v1%200_Final_6%20April%202016.pdf

United States Army Special Operations Command (no date). *"Little Green Men": A Primer on Modern Russian Unconventional Warfare, Ukraine 2013-2014*. Fort Bragg: United States Army Special Operations Command. Available from: www.jhuapl.edu/sites/defa ult/files/2022-12/ARIS_LittleGreenMen.pdf

United States Department of Defense, 2017. *Joint Publication 1-02 (JP 1-02), Dictionary of Military and Associated Terms*. Washington, DC: United States Department of Defense. Available from: https://apps.dtic.mil/sti/citations/AD1029823

United States Department of Defense, 2020. *Summary of the Irregular Warfare Annex to the National Defense Strategy*. Washington, DC: United States Department of Defense. Available from: https://media.defense.gov/2020/Oct/02/2002510472/-1/-1/0/Irregular-Warfare-Annex-to-the-National-Defense-Strategy-Summary.PDF

United States Department of the Treasury, 2021. Treasury Takes Further Action Against Russian-linked Actors. *United States Department of the Treasury*, Press Release, 11 January. Available from: https://home.treasury.gov/news/press-releases/sm1232

Vartanova, O., 2021. Russian Influence Operations (IPb) In 2021: Status and Expectations Through Geopolitical Assessment. *NSI*, 14 April. Available from: https://nsiteam.com/ russian-influence-operations-ipb-in-2021-status-and-expectations-through-geopolitical-assessment/

Votel, J., Cleveland, C. T., Connett, C. T., and Irwin, W., 2016. Unconventional Warfare in the Gray Zone. *Joint Forces Quarterly*, 80, 1st Quarter, 101–109. Available from: https:// ndupress.ndu.edu/Portals/68/Documents/jfq/jfq-80/jfq-80_101-109_Votel-et-al.pdf

Zhong, R., and Schuetze, C., 2021. Taiwan Wants German Vaccines. China May Be Standing in Its Way. *The New York Times*, 16 June. Available from: www.nytimes.com/2021/06/16/ business/taiwan-china-biontech-vaccine.html

12 Ubiquitous Technical Surveillance and the Challenges of Military Operations in the Era of Great Power Competition

Christopher Cruden

Introduction

Vigilant defensive management of our digital signatures and personal data emerges as a linchpin for the security of our forces and the triumph of future operations, particularly within the context of Great Power Competition. By employing identical methodologies to scrutinize adversaries' data, we gain the ability to preemptively identify their operations, leading to their eventual compromise and amplifying the impact of offensive information operations (OIE) and other kinetic or non-kinetic campaigns. Furthermore, surveying and analyzing our adversaries' porous data sources serves as a potent reconnaissance tool, enabling us to identify, develop, and exploit subsequent cyber operations.

Yet, persistent digital situational awareness embodies a double-edged sword. The collection and analysis of such "digital dust" bestows upon us an unwavering eye and an infinite memory, capable of yielding critical, actionable insights into adversary operations, personnel, and force movements. However, when our adversaries turn their unblinking eyes toward the U.S. military's past, present, and future activities, even our most elite forces, including Special Operations Forces (SOF), find their operational and technological advantages eroded.

To confront the realities of the digital threat environment and fortify ourselves for operations against near-peer adversaries, the U.S. military must acknowledge the imperative for corrective, protective, and proactive measures. Failing to adeptly manage our digital signatures will precipitate operational compromise, mission failure, and strategic setbacks in this new era of Great Power Competition. Above all, U.S. military leadership must comprehend the gravity of these challenges, adapt operational profiles accordingly, and prioritize decisive actions to safeguard our digital assets in the face of evolving threats.

On 26 October 2020, Philip Walton, a U.S. citizen living in Niger, was kidnapped from his farm by seven men armed with AK-47s and other weapons. Within three days, U.S. Navy SEALS successfully conducted a high-risk and immensely complicated mission to rescue Walton, executing a high altitude-low opening (HALO) parachute insertion onto the objective and killing six of the seven

DOI: 10.4324/9781003425304-13

kidnappers (Murphy 2018) before recovering Walton unharmed. This operation demonstrated the truly global reach of U.S. SOF and, most importantly, the speed with which the United States can and will act to protect its citizens abroad.

However, Walton's rescue – a tactical success at every level – was "outed" in near real-time by a Dutch aircraft spotting website, which provided live tracking of the operation using open-source software (Cenciotti 2020) and crowd-sourced data. Using tail numbers and live flight tracking apps, web sleuths unraveled the network of military and civilian aircraft that took part in the operation, exposing tactics, techniques, and procedures and jeopardizing future operational capabilities.

Walton's rescue was not the first-time sensitive SOF operations were revealed through the analysis of open sources, leaky apps, or user-generated data. In 2018, a data leak at the fitness tracking app (Hsu 2018) Strava compromised sensitive military base locations and patrol routes in Syria. But the quickness with which the operational security and secrecy of the Niger rescue mission unraveled demonstrates how SOF organizations have a digital signature security problem that needs to be addressed as they take the lead in the transition from counterterrorism to Great Power Competition. Conventional forces have the same problem – but at a far greater scale. It is not unreasonable to suspect that SOF and conventional forces' digital signatures are etched in foreign-held databases, gathered from more than twenty years of counterterrorism operations. It is also not unrealistic to expect that similar signatures have been compiled and stored for other operations, occurring outside the counterterrorism arena. Ultimately, we know that our near-peer adversaries are aware of the U.S. SOF digital signatures and are watching and waiting for them to appear in the "gray zones" (Singleton 2021) where the newest chapter of Great Power competition will play out.

SOF, Cyber, and Great Power Competition

Recognizing the superior performance of SOF in executing the counterterrorism fight for the last twenty years, policymakers have clearly articulated that SOF will have an integral role in executing operations below the level of armed conflict in this new era of Great Power Competition. As a result, commanders and decision-makers throughout the SOF community have emphasized building exquisite cyber capacity and the ability to conduct effective OIE – operations in the information environment. The new "door kickers" will be "coders" (Keller 2020) according to U.S. Special Operations Command (USSOCOM) commander General Richard Clarke. But regardless of whether they wield a keyboard, an MP-7, a Raspberry Pi, or a SCAR, the new generation of SOF operators will carry with them detectable digital signatures that can serve as early warning mechanisms (Tau 2021) of future operations for our adversaries. USSOCOM identified the digital signature issue in 2018, framing it as "signature management" (USSOCOM 2018) of no-fail missions, but failed to account for the type of digital signatures that leave no trace in the physical world. This was a major oversight – one that will not be overlooked by our Great Power Competitors.

As an example, a recent task force that was stood up in the Indo-Pacific (Pomerleau 2021) is charged with countering Chinese disinformation operations by establishing partnerships in the region and utilizing military information support operations. As the task force's partnerships mature and operations below the level of armed conflict increase in number and intensity, it is reasonable to anticipate an increase in the offensive nature of operations in both the cyber and physical environments. The Indo-Pacific mission also gives SOF new physical proximity to the Great Power Competition OIE battlespace. Our adversaries are aware and are watching to see if any familiar digital footprints pop up in their backyards.

While SOF will likely lead the way in adopting new technologies, training, and other mitigation measures, this problem set impacts the entire U.S. military. Great Power Competition and the next generation of operations will allow us to deny or undermine our adversaries' cyber and OIE advantages. It will also give our forces the ability to execute their own exclusive cyber capacities from unique platforms abroad. Unfortunately, we risk our newly tasked forces and the mission if we do not examine and modify our military's digital footprint before deployment into contested zones. This requires a thorough understanding of the constantly evolving digital world we live in – and the constant state of surveillance that exists there.

Ubiquitous Technical Surveillance

Ubiquitous technical surveillance (UTS) refers to the monitoring of individuals or groups using various technological means, such as cameras, microphones, and the hundreds of sensors we keep in our homes (Strange Sounds 2019) or that are in our phones. It has become increasingly prevalent (Lyon 2007) in modern society, with governments, businesses, and even individuals utilizing various forms of surveillance to monitor people's behavior, movements, and activities. The advent of modern technology has made it easier to collect and analyze vast amounts of data, including personal information and communication records. In many cases, this information is collected without the knowledge or consent of the individuals being monitored. While some forms of surveillance can be useful in detecting and preventing criminal activity, such as in the case of closed-circuit television (CCTV) monitoring in public spaces, the pervasive nature of UTS raises concerns (McCahill 2002) about privacy and civil liberties. Governments and law enforcement agencies have been among the most prominent users of surveillance technology. They have utilized various methods, including the use of facial recognition technology, to monitor people's activities in public spaces. This type of surveillance can track individuals attending protests and other political gatherings, raising concerns about the suppression of free speech and the right to assembly. Similarly, the collection of data from individuals' online activities has been used by governments to identify potential security threats and monitor the activities of dissidents and other individuals considered to be a threat to national security.

Critics argue that ubiquitous surveillance represents an unacceptable intrusion into individuals' private lives and undermines basic human rights, such as freedom of speech and the right to privacy (Sanchez 2009). In addition, there are concerns about the potential for such surveillance to be used for discriminatory purposes, such as racial profiling or political persecution. Despite these concerns, the use of UTS technology shows no signs of slowing down. According to Zuboff (2019), advances in artificial intelligence, machine learning, and other technologies are likely to make surveillance even more pervasive and sophisticated in the years to come. This new technological paradigm – the concept that we constantly shed "digital dust" all the more important as we prepare to confront near-peer adversaries with similar technical capabilities.

The fact is that military operations in an era of Great Power Competition require a thorough understanding of the digital signature of deployed troops as well as those in garrison. Understanding how a military unit looks in the context of UTS is the first step in ensuring better OPSEC for future operations. It is important to remember that UTS presents a major risk to all military operations – not just those conducted by SOF. And while SOF may lead the way in terms of technical innovation in mitigating UTS – it is necessary for all the military services to follow suit.

Digital Dust

The concept of "digital dust" (Kshetri 2018) refers to the unintentional trace data that individuals leave behind as they use digital technologies, such as emails, text messages, social media posts, and other digital services, often without realizing it. These bits of data can include things like search histories, social media activity, and location data, and can be used by companies to build detailed profiles of individuals and target them with advertising and other marketing tactics.

While this data may seem innocuous, it can be valuable to foreign militaries seeking to gain an advantage over the U.S. military.

Research on digital dust has highlighted the importance of privacy and data protection in the digital age. A study by the Pew Research Center (Madden and Rainie 2015) found that 91% of U.S. adults agreed that consumers have lost control over how their personal information is collected and used by companies.

The accumulation of digital dust by a technologically sophisticated adversary can have a tremendous impact on military operations. For example, an adversary can use digital dust to conduct advanced social engineering attacks. Sjouwerman (2023) argues that technically sophisticated social engineering attacks that involve using information gathered from public sources, such as social media posts in combination with artificial intelligence (AI), are tremendously effective. By using digital dust in this way, foreign militaries can increase the likelihood of their attacks being successful. Another way that foreign militaries can use digital dust is by using Open-Source Intelligence (OSINT) techniques. OSINT (NATO n.d.) refers to the collection, analysis, and dissemination of information from publicly available sources. By using OSINT techniques, foreign militaries can gather information about our military's personnel, equipment, and operations. This information can then be used to gain a better understanding of our capabilities and vulnerabilities, as well as to plan and execute (Kallberg 2022) cyber-attacks. For example, if a foreign military is able to identify a vulnerability in a particular piece of equipment used by the U.S. military, they could use that information to develop an exploit that would allow them to gain access to the equipment and potentially disrupt operations.

The implications of foreign militaries (NIST 2018) using digital dust against the U.S. military are significant. By using digital dust to gather information about our military's personnel, equipment, and operations, foreign militaries can gain a significant advantage in any potential conflict. They can use this information to plan attacks that are more likely to be successful and to identify vulnerabilities that can be exploited. This could put U.S. military personnel at risk and compromise the U.S. military's ability to carry out any number of operations.

The Surveillance Economy

The surveillance economy refers to the economic system that has emerged because of the extensive collection, analysis, and monetization of personal data through UTS technologies and practices. In this economy, individuals' personal information is treated as a valuable commodity that is collected by companies, often without participants explicit consent or knowledge, and used for targeted advertising, market research, and other purposes. Edward Snowden, a traitor though he may be, warned about the mass surveillance efforts being undertaken by corporations as well as our government back in 2013 (Hirst 2013). Simply put, while corporations use data gathered in the UTS environment to monetize digital dust, governments can use the same data to track and monitor individuals or groups that they want to investigate further.

One of the fundamental drivers of the surveillance economy is the rapid advancement of technologies, such as smartphones, social media platforms, and internet-connected devices – including standard military equipment. This ever-expanding web of technologies serve as the sensor backbone for the UTS threat environment. These technologies can generate vast amounts of data about individuals, including their online activities, preferences and according to Almeida, Shmarko and Lomas (2022) our most intimate behaviors. This data is collected by companies through various means, such as tracking cookies, device identifiers, and data brokers, and then processed and analyzed to create detailed profiles of individuals.

Companies involved in the surveillance economy range from technology giants like Google and Facebook to smaller data brokers that specialize in aggregating and selling personal information. These companies profit by leveraging individuals' data to deliver targeted advertisements, optimize user experiences, and develop products and services based on consumer insights. The more data they collect, the more accurate and valuable their profiles become, enabling them to offer highly personalized and effective marketing campaigns (Lauer and Lipartito 2021). The same tactics that advertisers use to build a pattern of life for potential customers are used by targeters across the intelligence community – albeit frequently with more access to data.

The surveillance economy and the ability to buy an inexhaustible amount of data have significant implications for military operations, both in terms of challenges and opportunities. On one hand, the vast amounts of personal data available through the surveillance economy can offer valuable intelligence and insights that can enhance military capabilities, including targeting of enemy leadership and conduct of information operations (IO).

An example of a significant challenge presented by the surveillance economy is that of the Cambridge Analytica scandal. During this time, the personal data of millions of Facebook users was harvested and used for political advertising purposes by Cambridge Analytica in the run-up to the 2016 U.S. presidential election. Though this was not carried out by a Great Power Competitor, it set the political atmosphere in the United States ablaze and exemplified the potential misuse of data in the surveillance economy (Lapowski 2018). It could be reasonably argued too that it created mistrust in basic American beliefs such as free elections and set the conditions for years of legal and criminal investigations. Were a Great Power Competitor to have been behind this sophisticated assault, it's reasonable to assess that they could be credited with one of the greatest influence operations in history. Now, imagine what a foreign intelligence service could do with such data to impact the deployment of U.S. troops to a certain region or country.

On the other hand, such data sets present significant military opportunities such as advanced and ultra-precise pattern of life development. In essence, access to this data is a targeter's dream. For example, the impact of near real-time location data for the leader of a terrorist organization would exponentially speed up the kill chain. In fact, Jackson alleged in 2011 that one of the reasons it took so long to target and dispatch Osama Bin Laden was his refusal to utilize technologies that he believed would give away his location. And while it could be argued that Bin Laden's lack of a digital signature was a signature in itself – it certainly kept him safe from the full collection and analytical might of the U.S. Intelligence Community.

Rightly or wrongly, privacy concerns regarding the U.S. government's purchase and use of such data create significant barriers to the U.S. governments' ability to participate in the surveillance economy – especially when U.S. persons data could be included (DOD Directive 3115.18). And while incidents such as the Cambridge Analytica scandal highlight the need for robust data protection regulations and ethical guidelines to ensure individuals' rights are respected and their data is not abused, the U.S. must realize that our competitors are not bound by such restrictions.

Digital Signature Management

Continuous collection of our digital signatures gives adversaries access to patterns and profiles that provide stunning insight into our operations, plans, and personnel. The constant stream of data emitted from our force's personal cellphones, laptops, and social media accounts – devices and platforms that intertwine our military's personal and professional lives – draws a distinct outline around troop movement and physical location and creates a trackable, traceable, and near-real-time record of activity. Further, implementing strictly disciplined avoidance of susceptible devices and applications is not only burdensome and limiting to both personnel and operations, but may in itself provide a significant indicator to adversaries in an environment where use of such devices and applications is the universal norm. Commanders and decision makers are now realizing this new domain applies globally, at all times, and for all of their personnel.

Advantages we currently maintain as critical to our operational success, such as human performance, overwhelming force, superior technology, phenomenal logistics, and insightful intelligence, are significantly degraded by ignorance of our own digital signatures. U.S. military operations will continue to be betrayed by poor digital hygiene unless military leadership addresses its digital signature problem now. And if the problem is not addressed across the military services, poor OPSEC – such as the intermingling of personal and operational electronic devices in training events and deployments – will contaminate the force's ability to mitigate the UTS environment in future operations. So, what can be done to secure U.S. military operations now and in the future?

Protect the Force, Preserve the Mission

Prohibiting our adversaries' ability to pull our digital dust from the ether and identify our personnel and operations within the white noise of the global data environment is potentially achievable. But it requires a new strategy. This new strategy to protect the force and preserve our operational advantage should be built on three pillars: digital signature awareness training, technical systems architecture, and compliance auditing.

Digital signature awareness training should support a baseline behavioral standardization for the force, to ensure understanding and best practices. Training must include periodic refreshers as technology changes, understanding of how technology supports and hinders operations, and the extent of the current and future threats posed by technology and how we use it. Training can also address specific aspects of exercises to add realism to the ubiquitous collection of data and associated threats, as well as simulating the OIE battlespace. Technology is constantly changing the operational environment and training should reflect the realities of the terrain. As members of the force become more educated on the threat environment, they will be able to adapt in response to unanticipated threats and navigate this terrain more quickly and securely.

Technical systems architecture begins with mission need and requires an understanding of what technologies are needed to accomplish the mission, what technologies will secure the mission, and how easily these technologies can scale securely. Teams should ensure that the architecture can democratize data and make it available to as many users as possible for internal and external analysis. Democratizing data serves two purposes: it demystifies the data itself; and it empowers personnel to develop tools, ask questions, and produce insights using various techniques. Ultimately, technology should support operations, not be a limiting factor. Additionally, as technology continues to evolve, it will require persistent monitoring by technical subject matter experts to incorporate new systems into existing infrastructure and phase out the older and more obsolete technologies.

Compliance auditing is necessary to ensure policies, training, and systems are protecting the force and advancing the mission. From another perspective, compliance auditing is really creating and fostering cyber situational awareness. Continuous monitoring for compliance is necessary to make better, more informed decisions as new technological threats to personnel and mission are identified. Over time, the auditing will be adopted as doctrine, providing great insight into internal processes that will reinforce the overall security of personnel and mission.

"Defend Forward"

Overall, the landscape of UTS, digital dust and the surveillance economy on military operations is navigable, but littered with potential pitfalls. While these technologies and practices offer invaluable intelligence opportunities, they also present risks to operational security, privacy, and significant ethical considerations for a Democratic society. Military organizations must carefully navigate these challenges by implementing robust cybersecurity measures, promoting digital hygiene practices, and developing policies that strike a balance between leveraging available intelligence and protecting operational security and privacy.

The constant defensive management of our digital signatures and personal data is critical for the security of the force as well as the success of future operations within the context of Great Power Competition. When we use the same methodologies to examine our adversaries' data, we will be able to identify their operations before they commence, leading to their compromise, and enhancing the effect of OIE and other kinetic or non-kinetic campaigns. Surveying and analyzing our adversaries' leaky data also acts as a reconnaissance tool that will lead to the identification, development, and exploitation of follow-on cyber operations.

Persistent digital situational awareness is a double-edged sword. Collection and analysis of such "digital dust" creates an unblinking eye and infinite memory that can provide key, targetable insights into adversary operations, personnel, and force movements. But when our adversaries turn their own unblinking eyes in the direction of the U.S. military's past, current, and future activities, even our most elite forces, such as SOF, lose operational and technological advantages.

The U.S. military must recognize the realities of the digital threat environment, how current operational profiles fit within it, and what continuing the status quo will mean for future operations against near-peer adversaries. Above all, U.S. military leadership must understand that failure to take corrective, protective, and proactive actions to manage their digital signatures will result in operational compromise, mission failure, and strategic loss in this new era of Great Power Competition.

References

Almeida D., Shmarko K., and Lomas E., 2022. The Ethics of Facial Recognition Technologies, Surveillance, and Accountability in an Age of Artificial Intelligence: a Comparative Analysis of US, EU, and UK Regulatory Frameworks. *AI Ethics*, 2(3): 377–387. DOI: 10.1007/s43681-021-00077-w

Cenciotti, D., 2020. Dissecting The U.S. Hostage Rescue Operation in Nigeria: Here Are All the Assets That Took Part In The Raid. *The Aviationist*, 9 November. Available from: https://theaviationist.com/2020/11/09/dissecting-u-s-hostage-rescue-operation-in-nigeria-here-are-all-the-assets-that-took-part-in-the-raid/

Hirst, M., 2013. The Conversation. Someone's Looking at You: Welcome to the Surveillance Economy, *The Conversation*, 26 July. Available from: https://theconversation.com/someones-looking-at-you-welcome-to-the-surveillance-economy-16357

Hsu, J., 2018. Strava Data Heat Maps Expose Military Base Locations Around the World. *Wired*, 29 January. Available from: www.wired.com/story/strava-heat-map-military-bases-fitness-trackers-privacy/

Kallberg, J., 2022. Open Data and OSINT as an Attack Vector. *Cyberwire*, 29 November. Available from: https://thecyberwire.com/stories/aa985dfcb7b446ba98124bf9d41ec45e/open-data-and-osint-as-an-attack-vector

Keller, J., 2020. SOCOM Chief: Door-Kickers Are Out, Cyber Operators Are. *In: Task & Purpose*, 12 May. Available from: https://taskandpurpose.com/news/special-operations-forces-cyber-warfare/

Kshetri, N., 2018. Blockchain's Roles in Meeting Key Supply Chain Management Objectives. *International Journal of Information Management*, 39, 80–89. https://doi.org/10.1016/j.ijinfomgt.2017.12.005

Lapowsky, I. 2018. Google Autocomplete Still Makes Vile Suggestions. *In: Wired*, 12 February. Available from: www.wired.com/story/google-autocomplete-vile-suggestions/

Lauer, J. and Lipartito, K., 2021. *Surveillance Capitalism in America*. Philadelphia, PA: University of Pennsylvania Press.

Lyon, D., 2007. *Surveillance Studies: An Overview*. New York: Cambridge Polity Press.

Madden, M. and Rainie, L., 2015. Americans' Attitudes About Privacy, Security and Surveillance. *Pew Research Center: Internet, Science & Tech*, 20 May. Available from: www.pewresearch.org/internet/2015/05/20/americans-attitudes-about-privacy-security-and-surveillance/

Mccahill, M., 2002. CCTV in Britain: The Growth of Surveillance. *Surveillance & Society*, 1(3), 272–292.

Murphy, J., 2018. *Special Forces HALO Parachuting: Go Big or Go Home*. *SOFREP*, 8 July. Available from: https://sofrep.com/news/special-forces-military-free-fall-go-big-go-home/

NATO, n.d. *NATO OSINT Handbook V 1.2*. Available from: https://archive.org/details/NATOOSINTHandbookV1.2/mode/2up

NIST, 2018. *Framework for Improving Critical Infrastructure Cybersecurity*. Version 1.1. Gaithersburg, MD: National Institute of Standards and Technology. https://doi.org/10.6028/nist.cswp.04162018

Pomerleau, M., 2021. Special Operations Team in Pacific will Confront Chinese Information Campaigns. *C4ISRNet*, 25 March. Available from: www.c4isrnet.com/information-warfare/2021/03/25/special-operations-team-in-pacific-will-confront-chinese-information-campaigns/

Sanchez, J., 2009. Security vs. Privacy, Reinterpreting the Fourth Amendment, *Ars Technica*, 11 November. Available from: https://arstechnica.com/tech-policy/2009/03/from-the-academy-the-end-of-privacy/

Singleton, C., 2021. Flip the Gray Zone Script: How the US Can Customize its Approach to China. *Defense News*, 26 February. Available from: www.defensenews.com/opinion/commentary/2021/02/26/flip-the-gray-zone-script-how-the-us-can-customize-its-approach-to-china/

Sjouwerman, S., 2023. Council Post: How AI Is Changing Social Engineering Forever. *Forbes*, 26 May. Available from: www.forbes.com/sites/forbestechcouncil/2023/05/26/how-ai-is-changing-social-engineering-forever/?sh=79a0daea321b

Strange Sounds, 2019. The Terrible Truth about Alexa: It is Made to Spy On Us. *Strange Sounds*, 5 May. Available from: https://strangesounds.org/2019/05/alexa-spy-device-amazon-video.html

Tau, B., 2021. The Ease of Tracking Mobile Phones of U.S. Soldiers in Hot Spots. *Wall Street Journal*, 26 April. Available from: www.wsj.com/articles/the-ease-of-tracking-mobile-phones-of-u-s-soldiers-in-hot-spots-11619429402

USSOCOM, 2018. *Signature Management (SIGMAN). Science and Technology: Preparing for the Future 2020-2030*. Available from: www.socom.mil/SOF-ATL/SOFHardProblemsDocumentLibrary/USSOCOM%20ST%20Signature%20Management%20-%20SIGMAN%20MegaTalker-01Mar18.pdf

Zuboff, S., 2019. *Age of Surveillance Capitalism: The Fight for a Human Future at the New Frontier of Power*. New York: Public Affairs.

13 Toward a Whole-of-Society Framework for Countering Disinformation

J.D. Maddox, Casi Gentzel, and Adela Levis

Introduction

Imagine the following scenario: A group of U.S. military and diplomatic officials meet to discuss a named operation. During the meeting, a heated argument erupts among a uniformed trio about the need to overcome the current doctrinal limits of information operations. After a few clever retorts from a diplomat about the primacy of public diplomacy, a visiting Silicon Valley technocrat chimes in, explaining how his one-click solution might work. The argument goes on, in a circular manner, as one side or another refers to definitions and authorities until, like many similar discussions, the meeting ends in an agreement to revisit the topic at another time. Frustratingly, in the time it took the officials to reach an unsatisfying conclusion, malign actors initiated an entirely new disinformation campaign to undermine trust in our democratic institutions and values.

We are constantly reminded of the real-world impact of disinformation, from Russia's long-standing active measures to weaken democratic and international institutions, to terrorist groups' deceptive recruitment tactics, to the impact of vaccine disinformation on COVID vaccine acceptance rates around the globe (Global Engagement Center 2020). Adding to the problem are deepfakes and other technological advances that are emerging as mainstream disinformation capabilities (Bradshaw and Howard 2019). Disinformation is a widespread and serious threat and countering it requires coordination and collaboration across a broad array of tools and actors to make meaningful progress.

The Department of Defense's 2014 doctrinal definition of information operations (IO) describes the objective to "influence, disrupt, corrupt, or usurp the decision-making of adversaries and potential adversaries while protecting our own" (Joint Chiefs of Staff 2014). However, in 2018, operations in the information environment (OIE) conceptually displaced IO (Joint Chiefs of Staff 2018). OIE attempts to "change or maintain the perceptions, attitudes, and other elements that drive desired behaviors of relevant actors." The U.S. Advisory Commission on Public Diplomacy defines public diplomacy (PD) as "activities intended to understand, inform, and influence foreign publics" (U.S. Advisory Commission on Public Diplomacy 2021). These definitions, and their ever-evolving supporting terms – informational power, perception management, psychological operations, strategic

DOI: 10.4324/9781003425304-14

communication, information warfare, influence operations – tend to confuse rather than clarify terms at a time when U.S. organizations are facing increasingly sophisticated and resolute adversaries in a complex and interconnected information environment.

Traditional U.S. definitions of IO and PD – implying distinct functions between the military and the diplomatic corps – *require our attention but should not hinder cooperative action* as the information environment becomes riskier due to technological advancements, steep competitor resource expenditures, and brazen adversarial operations. Our separate and distinct governmental authorities, our independent press, and our robust and freewheeling private sector and civil society are America's greatest strengths – they ensure the protections and the freedoms we tend to associate with being American. And to preserve these American values, we must ultimately be more efficient and effective than our authoritarian competitors who trade protections and freedoms for stronger control and speed of action.

Given the reality of decentralized and underfunded U.S. activity in the counter-disinformation sphere, the community of practice needs to shelve the dictionary and begin sharing functionality among themselves and with partners while simultaneously fostering trust within the community and maintaining compliance with the law. More useful than schoolhouse definitions – and key to enabling coordinated impact – is a comprehensive understanding of the available tools across the full spectrum of actors and functions and agreement on mutually beneficial roles. In short, the authors propose a whole-of-society framework, and this chapter works toward identifying roles and functions that are available for a comprehensive whole-of-society approach.

This chapter proposes a framework and outlines capabilities that are focused on the shared objective of countering disinformation that undermines U.S. policies, stability, and national security. The term "countering disinformation," a term that is almost as misunderstood and redefined as IO and PD, does not consist merely of counter-messaging but also of *proactive measures* that use facts to inform audiences, reduce the impact of disinformation, and promote freedom of expression – activities that can be functionally categorized under communication, resilience, disruption, and regulation.

Communication

Proactive communication, setting the narrative, or filling the communication void before mis- or disinformation can distort the truth is a critical undertaking that requires understanding and implementation of the full spectrum of communication capabilities. Increasing transparency and building trust in democratic values and institutions is the key to success. While counter-messaging should not be the first resort, it does have its time and a place alongside broader, proactive communication efforts.

The United States, as a society and nation, has an extraordinarily broad range of disparate communicators and voices, ranging from governmental institutions to the private sector to civil society actors and organizations. Communication platforms are also constantly evolving and expanding their reach as the production and consumption of social media, print and digital media, radio, and television continuously grow and change. Within the U.S. government alone, communication encompasses several activities:

- **Public Affairs**: Experienced public affairs professionals adeptly navigate owned, paid, and earned media, building the necessary relationships to secure ideal placement of targeted content to inform audience attitude or behavior (Joint Chiefs of Staff 2016). Standard public affairs tools include press briefings, interviews, press releases, media flyovers, media co-ops, advertising campaigns, public events, and branding. Techniques such as narrative, framing, and endorsement are a mainstay in this space.
- **Key Leader Engagements**: Key leader, diplomatic, or other official engagements are instruments for impactful communication to specific audiences (Burke and Connell 2020). The tools in this category, more so than the method of engagement or the talking points, are the communicators themselves, because what is said is often less important than how it is said or who says it.
- **Strategic Signals**: The military also communicates through action; in fact, considering its prominence, the U.S. military sends messages both through action as well as inaction. Strategic signals frequently include freedom of navigation operations, capability demonstrations, exercises, investments, and asset placement, which can, more or less subtly, communicate resolve, commitment, and priority, or project battlefield superiority (Department of Defense 2019).
- **Public Diplomacy**: Public diplomacy engagements and communications can vary in style and execution based on the country of execution and can include speaker series, social and traditional media campaigns, American Spaces, exhibitions, and seminars (additional public diplomacy tools are categorized under resilience) (Department of State 2020).
- **Military Information Support Operations** (MISO): One element of doctrinal IO, MISO offers a broad range of communication opportunities dependent upon the environment, including online or printed magazines or journals, websites, news apps, radio programs, leaflets, WebOps, and text messages (Joint Chiefs of Staff 2014).
- **Counter Messaging**: The use of this tool assumes the disinformation message is already out. Once spread, disinformation can be difficult to counter; however, one of the more successful counter-messaging tactics is exposure – exposing the source and deceptive or malign nature of the message as opposed to focusing on the details of its inaccuracy (Global Engagement Center 2020). Other tactics include denigration, boycotting, and, under specific limited circumstances, deception.

Resilience

Building resilience, reducing vulnerability, and inoculating populations against the effects of mis- and disinformation are a few of the most effective ways to counter disinformation and are an absolute necessity in the current global information environment (Braddock 2019). Building resilience to disinformation involves the producers of information, education systems, media associations, public diplomacy practitioners, and organizations equipping and empowering society to recognize how disinformation is created and spread and to understand its impact. It encourages individuals and organizations to think critically about how we consume and share information and creates protective barriers to disinformation's negative effects on society. Responsible digital citizenship comes down to the individual, but this skill can be fostered through the following measures:

- **Digital literacy**: It is vital to integrate media and digital literacy and updated critical thinking skills into educational curricula, adult education and professional development opportunities, and public service announcements or other educational ventures. Games can be effectively used to inoculate populations against the techniques and methods used by malicious actors and teach populations to recognize disinformation when they see it (Roozenbeek and van der Linden 2020). This is not a quick fix, but is critical to a long-term solution.
- **Civics**: A renewed effort to broaden and strengthen pro-democratic values and civic education through various educational and other civil society organizations is also necessary. These efforts should incorporate skills for understanding information related to government and election processes and focus on America's pluralistic history, reasoning, and critical thinking.
- **Journalism**: Media plays a critical role in a society's resilience. It is critical to promote journalistic standards to safeguard independent, fact-based, investigative journalism and establish and bolster fact-checking standards and norms (International Fact Checking Network 2022).
- **Public diplomacy**: Public diplomacy efforts can build resilience to disinformation in foreign populations through a myriad of programs, which include facilitating professional, cultural, and educational exchanges, supporting English language education, training journalists, conducting capacity-building workshops, supporting local credible voices, and supporting young leaders' initiatives.

Disruption

A more pointed approach leverages technology to stem the flow of disinformation at its source or as it is in transit to the consumer. These technical solutions are often part of the toolbox used by the broader tech sector and internet platforms, while some capabilities are used by law enforcement and the military overseas, including in the execution of cyberspace operations. The most common technical solutions include:

- **Validation**: Identity, content, and site authentication are some of the tools that can support the validation of social media, websites, political ads, and more that are increasingly subject to disinformation. Current law, Section 230 of the

Communications Decency Act, places the responsibility for this kind of work on the internet platforms, but new tools – such as tools to track who created, updated, deleted, and read information – have potentially broader applications when used in a rights-respecting manner (Cornell Law School 2019).

• **Blocking**: Techniques such as content moderation and algorithmic filtering of content are social media companies' core means of keeping unwanted content from infecting the information environment (Twitter 2020). These techniques can be controversial, as some see them as censorship, though internet platforms create and enforce their own terms of service.

• **Destruction**: Destruction of adversarial capabilities is a high-risk, high-reward calculation. The military's destruction of centers of gravity, including ISIS's media production capabilities and media dissemination sites, for example, yielded at least short-term gains against a flexible media system.

• **Cyberspace operations**: Another high-risk, high-reward tactic, , cyber operations and hacking and dumping offer temporary disruption of sites, servers, and communication systems, and serve as a clear message to adversaries that they are vulnerable and that the United States has the advantage. Again, the military's cyber disruption of ISIS's online operations is a good example (Temple-Raston 2019).

• **Enforcement**: Recent arrests, indictments, and sanctions of foreign agents involved in disinformation operations against U.S. audiences have served to send a clear message (Department of Justice 2018a,b; Department of Treasury 2020). Federal law enforcement signals the United States' resolve every time it takes action.

Regulation

Because disinformation will never disappear, regulation is a necessary component of any proposed approach to counter the impact of disinformation directed at the United States. Effective regulation should source input from local and national legislators, media associations, internet platforms and the broader tech sector, and international organizations. Several approaches to regulation are required and should include the following:

• **Legislation**: The U.S. Congress and state legislatures have taken a crucial role in managing the United States information environment. Recent bills such as SAFE TECH, Honest Ads, and ACCESS demonstrate legislators' new willingness to participate in this complex issue (Bilirakis 2018; Klobuchar 2019; Schakowsky 2020); however, partisan disagreement stymies enactment of new bills and recent flawed foreign legislative models set bad precedents.

• **Regulation**: American TV networks and journalistic organizations have long maintained standards for information and reporting acceptability. The FCC once attempted to enforce honest coverage through the fairness doctrine, which was abandoned due to its perceived effect of "chilling" free speech. Similar models have been discussed and suggested for the internet.

- **International cooperation**: Diplomatic efforts to counter the effects of disinformation require extraordinary sophistication of coordination but can result in strategic moral gains against adversaries.

Weaving It All Together

Categorizing activities to counter disinformation within the functions of communication, resilience, disruption, and regulation supports a focus on outcomes rather than creating greater bureaucratic division. Opening the information aperture to consider a broader spectrum of actors and capabilities allows communicators, strategists, and policymakers to construct impact-based activities by combining disparate capabilities. No single agency and no single tactic is capable of countering disinformation on its own. Therefore, we must be committed to learning, collaborating, and innovating.

Countering disinformation requires *creative, collaborative, and coordinated* implementation of the full spectrum of activities to achieve the greatest impact. To that end, the U.S. Congress has mandated the Global Engagement Center (GEC) to direct, lead, synchronize, integrate, and coordinate efforts of the federal government to recognize, understand, expose, and counter foreign state and foreign nonstate actors' propaganda and disinformation efforts.

U.S. departments and agencies must continue to undertake counterdisinformation activities in compliance with their existing authorities – including limits on domestic influence operations, for example – to preserve our democratic values and institutions and reverse the erosive effects of disinformation on society while promoting freedom of expression. But by reorienting planning efforts to capitalize on whole-of-government and whole-of-society functions and strengths, the United States can gain new advantages over adversaries and competitors who leverage strong control and faster adoption of novel influence techniques to undermine our social fabric.

A Real-Time Ukraine Case Study

The Government of Ukraine's rapid information operations (IO) reaction to the Russian invasion beginning on 24 February 2022 offers important lessons for the development of whole-of-society IO solutions, especially highlighting new technological requirements for national-level efforts. The following case study assesses activities for the first month of Ukrainian defenses, offering a realistic look at how a nation under duress has implemented and unified sometimes disparate capabilities. Ukraine's integration of novel technological capabilities, including decentralized nongovernment capabilities, in combination with more traditional strategic communications capabilities, demonstrates the organizational flexibility necessary to meet acute threats.

"Digital Militias" Take Up Arms against Russia

The emergence of private militia-style movements online seems to have been unprecedented in its scope and importance to the overall effort of the Government of Ukraine, and especially effective for unifying IO actors against the Russian invasion. The Government of Ukraine did not hesitate to incorporate private groups, opening channels of communication with these groups as they undertook significant operations that otherwise might have been considered the exclusive roles of government organizations. The distinguishing feature of Ukraine's IO efforts from the earliest moments of the conflict was its inclusion of a broad array of independently-originated private militia-style IO functions, including centralized information operations, as well as both offensive and defensive IO-supportive cyber operations.

- **Hub-and-spoke messaging**: The Government of Ukraine's Centre for Strategic Communications and Information Security, under the Ministry of Culture and Information Policy of Ukraine, performed a core function of developing original content, redistributing existing content, and centralizing information awareness (Stratcom Centre UA 2022). The @StratcomCentre page on Twitter (now X) functioned as a reliable repository for IO content, which was routinely redistributed by pro-Ukraine actors.
- **Digital activists**: Civil society actors such as Bellingcat actively crowdsourced and otherwise conducted rapid digital investigations of Russian atrocities and other violations of civil and military law, attracting Western attention to issues that might otherwise have gone uninvestigated (Basu 2022). These digital investigations would be included on various pro-Ukraine IO channels.
- **Volunteer cyber warriors**: The militia-style IT Army of Ukraine formed on the communications apps Telegram and Discord as the invasion took place, attracting over 350,000 global followers in the first month, an untold number of whom participated in its operations. The IT Army of Ukraine used its Telegram channel to centralize targeting of Russian and pro-Russian websites and other digital venues by its volunteer cybersecurity activists (Stokel-Walker and Milmo 2022). The group identifies targets in real-time, such as Russian banks and military facilities for Distributed Denial of Service attacks and publishes the results of its cyber-attacks on a rolling basis, despite the openness of the platform to pro-Russia observers. The efforts of the IT Army of Ukraine were complemented by activities by the hacktivist group Anonymous, who made a highly visible announcement of their operations against Russia (#OperationRussia) immediately after the invasion, and hacked and dumped data from the Russian Central Bank in one high-profile operation (Pitrelli 2022). Additionally, bug bounty groups initiated crowdsourced efforts to find and attack Russian cyber vulnerabilities (Haynes 2022).
- **Crypto donors**: The Government of Ukraine quickly established a cryptocurrency funding mechanism for international donations to the fight against Russia. Ukraine initially used the Kuna, FTX, and Everstake cryptocurrency exchange systems to enable Bitcoin, Ethercoin and other crypto denominations to be transferred to the Government of Ukraine (Wright 2022). Ukraine's Deputy Minister of Digital Transformation claimed that Ukraine had received "close to $100 million" in crypto donations after two weeks – an unverifiable sum (Amitoj 2022).

- **Guerilla journalists**: Independent journalists served a core IO function by providing on-the-ground reports of Ukrainian defiance, civilian commitment to the
fight, defeats of Russian military units, as well as atrocities committed by Russian
military units – all of which were rapidly repeated in Western media (Associated
Press 2022; Melchior 2022; Ponomarenko 2022; Amnesty International 2022).
Noticeably absent was regular coverage of front-line infantry engagements.
Among journalistic platforms, the Kyiv Independent stood out as a primary
source, with independent journalist Illia Ponomarenko (@IAPonomarenko) – a
self-styled "guerilla" – serving as a prolific and influential war correspondent
on social media. These reports appeared to correspond to a core narrative of
Ukrainian unity and defiance despite overwhelming odds.

Mainstream Social Media Platforms Offer Partial Solutions to Russian Disinformation

The Government of Ukraine and Western activists notably pressured social media
platforms with a presence in Ukraine and Russia to act on Ukraine's behalf by
stemming anti-Ukraine disinformation and propaganda, limiting advertisement-
based financial incentives for Russia-based companies and government
organizations, and modifying the algorithmic prioritization of social media content
(Human Rights Watch 2022). The efforts of these social media platforms came
only after the fact of the Russian invasion, remained piecemeal in their application,
and were notably resisted by many other platforms.

- After receiving pressure, platforms such as Meta/Facebook, Snap, and YouTube
demonetized some Russian channels (Ramishah 2022).
- After the invasion initiated, Meta/Facebook, Instagram, Twitter/X, and YouTube
restricted posts and accounts run by Russian state media (Bond 2022).
- Also, after the invasion, some platforms like Meta/Facebook moved to modify
their content prioritization algorithms (Haugen 2022).
- Meta/Facebook and Twitter/X also started labeling posts that include links
to Russian state-backed media outlets so people know what they're reading
(Meta 2022).

IO Operations – A Core Competency of Ukrainian Defense: While integrating
novel IO capabilities, the Government of Ukraine also undertook a more traditional centralized campaign that included IO narratives from senior leadership and
that pulled from more traditional IO resources that had already been part of the
Government of Ukraine's capabilities.

- **Presidential appeals to emotion**: Perhaps the most visible of all IO initiatives
by the Government of Ukraine, and one that began before the 24 February invasion, is the implementation of daily live and recorded briefings by President
Volodymyr Zelenskyy on social media (Zelenskyy 2022). These briefings

included persistent narratives such as condemning the Russian invasion, expressing unity of the Ukrainian people against an unprovoked invasion, goading the West into providing support, appealing to the Russian people to defy Vladimir Putin, and providing evidence of Russian atrocities – especially the deaths of children. The tone of these daily briefings generally evolved in the first few weeks from demonstrations of Ukrainian resolve to appeals for global support.

- **A focus on western audiences**: From the U.S. perspective, the culmination of President Zelenskyy's online IO efforts took place on 16 March 2022, when he directly addressed the U.S. Congress during a public transmission (CNN 2022). The 16 March address reemphasized Ukraine's core unity and defiance narratives but made a direct appeal to the U.S. Government for support, and especially for a no-fly zone over Ukraine. The address was accompanied by numerous graphic images – especially images of the murders of Ukrainian children and adult civilians by Russian forces – and was received with a standing ovation by the U.S. Congress. Additional Western-focused messaging efforts of special interest included expressions of fear that Russia would imminently use chemical or nuclear weapons, or would enable a "meltdown" at the Chernobyl nuclear power plant – efforts that seemed calibrated to draw a U.S. response (Schnell 2022; Sukhov 2022; Sale 2022).

- **All-source imagery/content integration**: The imagery and reporting integrated into these open briefings and other media efforts was selected from Government of Ukraine sources and a broad array of private media sources who were freely active in Ukraine, sometimes collocated with Ukrainian defensive forces. Ukraine's Special Operations Forces included four designated Psychological Operations units regionally assigned, and they are probably the originators of drone footage and on-the-ground counterattack footage that was used to demonstrate Ukrainian successes against Russia, but sourcing is scant. Additionally, Western media were among the frontline war correspondents in Ukraine, with at least four U.S. journalists dying in the performance of their jobs in the first weeks of the fight.

A War of Images

Among the prominent Ukraine IO imagery-based campaigns emerging on social media since the 24 February invasion were those that demonstrated the defiance of the Ukrainian people and army, and the abhorrence of civilian deaths. Some of the original images or information was attributed to official Government of Ukraine sources, but the attribution was not always clear.

- **Russian defeat statistics**: A daily tally of Ukrainian military defeats of Russian invading troops and equipment was regularly posted to official and unofficial social media channels, attributed to the Ukraine Ministry of Defense (Query 2022). These statistics could not be verified by Western authorities but were cited frequently by commentators (Figure 13.1).

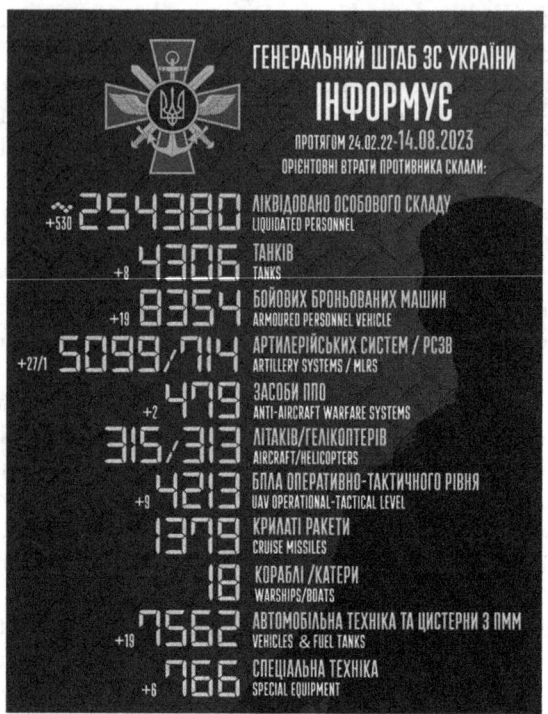

Figure 13.1 A Daily Tally of Russian Losses Produced by the Ukraine Ministry of Defence.

Source: Ukraine Ministry of Defence (2023).

- **Snake Island defiance legend**: Shortly after the invasion, according to a story sometimes attributed to the Ukraine Defense Ministry, a group of 13 Ukrainian border troops on Snake Island were killed by a Russian naval bombing, but only after a final radio transmission in which the troops defiantly said, "go f*ck yourself" (*idi nakhui*). The phrase was quickly disseminated online and turned into a viral meme (Walker 2022) and eventually a postage stamp (see Figure 13.2), although the original story would be disputed.
- **Tractors pulling military equipment**: A few weeks into the invasion, an image of a Ukrainian farmer pulling a confiscated Russian tank emerged on social media, and it was quickly turned into a viral meme, with following versions mockingly showing a tractor pulling a Russian missile system, a Russian ship, etc. (Daily Mail 2022). The image would later be turned into a postage stamp by the Government of Ukraine.
- **Drone footage of destroyed Russian equipment and forces**: Pro-Ukraine reporting regularly included drone imagery (see Figure 13.3) showing the destruction of Russian equipment, including missile systems, tanks, armored

Figure 13.2 A Postage Stamp Showing a Ukrainian Soldier Giving the Middle Finger to the Ship.

Source: Groh (2022).

Figure 13.3 Picture from Drone Footage of Destroyed Russian Equipment.

Source: Kamikaze Drone Attack (1945, 2022).

personnel carriers, and more (Khan 2022). Footage did not normally include graphic displays of Russian deaths.

- **Russian atrocities against civilians**: As early as 27 February, European news media alleged Russian responsibility for the deaths of Ukrainian civilians, and especially children. Among the prominent images distributed was a photo by AP photographer Evgeniy Maloletka showing a dying child in Mariupol (Graham-Harrison 2022). The Ministry of Internal Affairs of Ukraine distributed additional images as evidence of Russian atrocities.

The Galvanizing Effects of an Existential Threat

One key lesson learned from Ukraine's early IO response to the Russian invasion of 2022 is that a system that rapidly incorporates unconventional IO "force multipliers" – such as Ukraine's acceptance of private digital activists against Russian targets – may exceed its normal and expected capacity for IO activities. Ukraine's flexibility may be a replicable model for open societies' whole-of-society IO defenses and demands consideration of how these fast-moving nongovernmental functions can be integrated into government systems that are normally hierarchical and that typically require lengthy processes to permit even the most basic IO activities. One additional potential limitation on the incorporation of these nongovernmental functions is the security risk they represent – especially the risk of exposing imminent security operations to sabotage, like the risks inherent to the publication of Russian cyber targets by the IT Army of Ukraine on Telegram. The Government of Ukraine seems to have calculated that the reward of unconventional capabilities outweighs the risk in the face of the existential threat posed by the Russian invasion.

Among the whole-of-society IO categories described in this chapter, it seems that the Government of Ukraine relied heavily on disruption capabilities in the early response to Russia's invasion. These included blocking, destruction, and cyberspace operations, but these were complementary of a structured central communications coordination effort, which strongly featured key leader engagement, public diplomacy, Ukraine-equivalent MISO capabilities, and counter messaging. While these features of the Ukraine IO effort are likely driven by the unique conditions of Russia's invasion, we can assume that if under duress, the United States would also move quickly to integrate innovative capabilities. War, after all, is the mother of innovation.

References

Amitoj, S., 2022. Ukraine Has Received Close to $100M in Crypto Donations. *CoinDesk*, 9 March. Available from: www.coindesk.com/business/2022/03/09/ukraine-has-received-close-to-100-million-in-crypto-donations/ [Accessed 28 March 2022].

Amnesty International, 2022. Russian Military Commits Indiscriminate Attacks During the Invasion of Ukraine. *Amnesty International*, 25 February. Available from: www.amnesty.org/en/latest/news/2022/02/russian-military-commits-indiscriminate-attacks-during-the-invasion-of-ukraine/ [Accessed 28 March 2022].

Associated Press, 2022. AP Photos: Day 17: Images of Destruction, Ukrainian Defiance. *Associated Press*, 12 March. Available from: https://apnews.com/article/russia-ukraine-europe-666be0a0cd62580b920452338e4c4de8 [Accessed 28 March 2022].

Basu, T., 2022. The Online Volunteers Hunting for War Crimes in Ukraine. *MIT Technology Review*, 16 March. Available from www.technologyreview.com/2022/03/16/1047322/ukraine-russia-war-crimes-evidence/ [Accessed 28 March 2022].

Bilirakis, G., 2018. *H.R.4561 – SAFE TECH Act*. Washington, DC: United States Congress. Available from: www.congress.gov/bill/115th-congress/house-bill/4561 [Accessed 28 March 2022].

Bond, S., 2022. Facebook, Google and Twitter Limit Ads over Russia's Invasion of Ukraine. *National Public Radio*, 26 February. Available from: www.npr.org/2022/02/26/1083291 122/russia-ukraine-facebook-google-youtube-twitter [Accessed 28 March 2022].

Braddock, K., 2019. Vaccinating Against Hate: Using Attitudinal Inoculation to Confer Resistance to Persuasion by Extremist Propaganda. *Terrorism and Political Violence*, 34(2), 240–262. https://doi.org/10.1080/09546553.2019.1693370.

Bradshaw, S. and Howard, P. N., 2019. *The Global Disinformation Disorder: 2019 Global Inventory of Organised Social Media Manipulation*. Working Paper 2019.2. Oxford, UK: Project on Computational Propaganda. Available from: https://demtech.oii.ox.ac.uk/wp-content/uploads/sites/93/2019/09/CyberTroop-Report19.pdf [Access 28 March 2022].

Burke, J. and Connell, T., 2020. PACOM Key Leader Engagement Process. *In: Proceedings of the 2020 Annual General Donald R. Keith Memorial Capstone Conference: A Regional Conference of the Society for Industrial and Systems Engineering*, 30 April 2020 West Point, New York, USA. Available from: www.ieworldconference.org/content/WP2020/Papers/GDRKMCC_20_53.pdf [Accessed 28 March 2022].

CNN, 2022. Zelensky Gets Standing Ovation After Speech to Congress. *YouTube*, 16 March. Available from: www.youtube.com/watch?v=RZ_ykL6QJE0 [Accessed 28 March 2022].

Cornell Law School, 2019. *47 U.S. Code § 230 – Protection for Private Blocking and Screening of Offensive Material*. Ithaca, NY: Legal Information Institute. Available from: www.law.cornell.edu/uscode/text/47/230 [Accessed 28 March 2022].

Daily Mail, 2022. Video: Ukrainian Farmer Appears to Tow Away a Russian Tank. *Daily Mail*, 28 February. Available from: www.dailymail.co.uk/video/news/video-2630 689/Video-Ukrainian-farmer-appears-tow-away-Russian-tank.html [Accessed 28 March 2022].

Department of Defense, 2019. *Report to Congress: Annual Freedom of Navigation Report*. Washington, DC: Department of Defense. Available from: https://policy.defense.gov/Portals/11/Documents/FY19%20DoD%20FON%20Report%20FINAL.pdf?ver=2020-07-14-140514-643&timestamp=1594749943344 [Accessed 28 March 2022].

Department of Justice, 2018a. *Russian National Charged in Conspiracy to Act as an Agent of the Russian Federation Within the United States*. Washington, DC: DOJ Office of Public Affairs. Available from: www.justice.gov/opa/pr/russian-national-charged-conspiracy-act-agent-russian-federation-within-united-states [Accessed 28 March 2022].

Department of Justice, 2018b. *Internet Research Agency Indictment*. Washington, DC: Department of Justice. Available from: www.justice.gov/file/1035477/download [Accessed 28 March 2022].

Department of State, 2020. *American Spaces Handbook*. Washington, DC: Department of State. Available from: https://americanspaces.state.gov/managing-your-space/american-spaces-handbook/ [Accessed 28 March 2022].

Department of Treasury, 2020. *Treasury Increases Pressure on Russian Financier*. Washington, DC: Department of Treasury. Available from: https://home.treasury.gov/news/press-releases/sm1133 [Accessed 28 March 2022].

Global Engagement Center, 2020. *GEC Special Report: Pillars of Russia's Disinformation and Propaganda Ecosystem*. Washington, DC: Department of State. Available from: www.state.gov/wp-content/uploads/2020/08/Pillars-of-Russia%E2%80%99s-Disinformation-and-Propaganda-Ecosystem_08-04-20.pdf [Accessed 28 March 2022].

Graham-Harrison, E., 2022. "It's Stomach-Turning": The Children Caught Up in Ukraine War. *The Guardian*, 27 March. Available from: www.theguardian.com/world/2022/feb/27/children-caught-up-in-ukraine-war [Accessed 28 March 2022].

Groh, Boris, 2022. Марка «Русскій воєнний корабль, іди…! Слава Україні!». Stamp itself [online], https://commons.wikimedia.org/w/index.php?curid=116873457 [Accessed 5 July 2023].

Haugen, F., 2022. Nick Clegg Has the Power Now to Right Facebook's Wrongs. This is How He Should Do It. *The Guardian*, 2 March. Available from: www.theguard ian.com/commentisfree/2022/mar/02/nick-clegg-facebook-wrongs [Accessed 28 March 2022].

Haynes. A., 2022. Crowdsourced Security is Out of Control in the Ukraine Conflict. *United States Cybersecurity Magazine*. Available from www.uscybersecurity.net/crowdsourced-security-out-of-control-in-the-ukraine-conflict/ [Accessed 28 March 2022].

Human Rights Watch, 2022. Russia, Ukraine, and Social Media and Messaging Apps. *Human Rights Watch*, 16 March. Available from: www.hrw.org/news/2022/03/16/russia-ukraine-and-social-media-and-messaging-apps# [Accessed 28 March 2022].

International Fact Checking Network, 2022. *Commit to Transparency – Sign Up for the International Fact-Checking Network's Code of Principles*. St. Petersburg, FL: Poynter. Available from: www.ifcncodeofprinciples.poynter.org/ [Accessed 28 March 2022].

Joint Chiefs of Staff, 2014. *Joint Publication 3-13: Information Operations*. Washington, DC: Department of Defense. Available from: www.jcs.mil/Portals/36/Documents/Doctr ine/pubs/jp3_13.pdf [Accessed 28 March 2022].

Joint Chiefs of Staff, 2016. *Joint Publication 3-61: Public Affairs*. Washington, DC: Department of Defense. Available from: www.jcs.mil/Portals/36/Documents/Doctr ine/pubs/jp3_61.pdf [Accessed 28 March 2022].

Joint Chiefs of Staff, 2018. *Joint Concept for Operating in the Information Environment (JCOIE)*. Washington, DC: Department of Defense. Available from: www.jcs.mil/ Portals/36/Documents/Doctrine/concepts/joint_concepts_jcoie.pdf [Accessed 28 March 2022].

Kamikaze Drone Attack. 1945, 2022 [online]. www.msn.com/en-us/news/world/kamikaze-chaos-new-ukraine-footage-shows-how-drones-can-kill-tanks/ar-AA1dg8ZB [Accessed 5 July 2023].

Khan, L., 2022. How Ukraine Is Using Drones Against Russia. *Council on Foreign Relations*, 2 March. Available from: www.cfr.org/in-brief/how-ukraine-using-drones-against-russia [Accessed 28 March 2022].

Klobuchar, A., 2019. *S.1356 – Honest Ads Act*. Washington, DC: United States Congress. Available from www.congress.gov/bill/116th-congress/senate-bill/1356/text?module= bill&controller=bill&format=txt [Accessed 28 March 2022].

Melchior, J., 2022. Civilians Prepare to Defend Ukraine. *Wall Street Journal*, 30 January. Available from: www.wsj.com/articles/civilians-prepare-to-defend-ukraine-russia-attack-training-war-putin-kyiv-military-11643566341 [Accessed 28 March 2022].

Meta, 2022. Meta's Ongoing Efforts Regarding Russia's Invasion of Ukraine. *Meta/ Facebook*, 26 February. Available from: https://about.fb.com/news/2022/02/metas-ongo ing-efforts-regarding-russias-invasion-of-ukraine/ [Accessed 28 March 2022].

Pitrelli, M., 2022. Anonymous Declared a "Cyber War" Against Russia. Here are the Results. *CNBC*, 16 March [online]. Available from: www.cnbc.com/2022/03/16/what-has-anonym ous-done-to-russia-here-are-the-results-.html [Accessed 28 March 2022].

Ponomarenko, I., 2022. EXCLUSIVE: Voice Message Reveals Russian Military Unit's Catastrophic Losses in Ukraine. *Kyiv Independent*, 2 March. Available from: https://kyiv independent.com/national/exclusive-voice-message-reveals-russian-military-units-catas trophic-losses-in-ukraine/ [Accessed 28 March 2022].

Query, A., 2022. Russia Faces Heavy Losses as It Attacks Ukraine on All Fronts. *Kyiv Independent*, 29 February [online]. Available from: https://kyivindependent.com/ national/russia-faces-heavy-losses-as-it-attacks-ukraine-on-all-fronts/ [Accessed 28 March 2022].

Ramishah, M., 2022. Meta and YouTube Block Russian State Media from Monetizing on Its Platforms. *CNN*, 26 February. Available from: www.cnn.com/2022/02/26/tech/meta-yout ube-facebook-rt-demonetize/index.html [Accessed 28 March 2022].

Roozenbeek, J., and van der Linden, S., 2020. Breaking Harmony Square: A Game that "Inoculates" Against Political Misinformation. *Harvard Kennedy School (HKS) Misinformation Review*. Available from: https://doi.org/10.37016/mr-2020-47 [Accessed 28 March 2022].

Sale, I., 2022. Ukraine War: Chernobyl Radiation Fears as Minister Calls for Russia to Allow for Urgent Repairs. *SkyNews*, 9 March. Available from: https://news.sky.com/story/ukra ine-war-chernobyl-radiation-fears-as-minister-calls-for-russia-to-allow-for-urgent-repa irs-12561615 [Accessed 28 March 2022].

Schakowsky, J., 2020. *H.R.6487 – ACCESS Act*. Washington, DC: United States Congress. Available from www.congress.gov/bill/116th-congress/house-bill/6487 [Accessed 28 March 2022].

Schnell, M., 2022. US, Ukrainian Officials Brace for Possible Russian Chemical Attack. *The Hill*, 13 March. Available from: https://thehill.com/homenews/sunday-talk-shows/ 598030-us-ukrainian-officials-brace-for-possible-russian-chemical-attack [Accessed 28 March 2022].

Stokel-Walker, C. and Milmo, D., 2022. "It's the Right Thing to Do": The 300,000 Volunteer Hackers Coming together to Fight Russia. *The Guardian*, 15 March. Available from: www.theguardian.com/world/2022/mar/15/volunteer-hackers-fight-russia [Accessed 28 March 2022].

Stratcom Centre UA, 2022. The Centre for Strategic Communications and Information Security under the Ministry of Culture and Information Policy of Ukraine. *Government of Ukraine*. Available from https://stratcomcentreua.medium.com/about [Accessed 28 March 2022].

Sukhov, O., 2022. Is Putin Going to Launch a Nuclear War? *Kyiv Independent*, 18 March. Available from: https://kyivindependent.com/national/analysis-is-putin-going-to-launch- a-nuclear-war/ [Accessed 28 March 2022].

Temple-Raston, D., 2019. How the US Hacked ISIS. *National Public Radio*, 26 September. Available from: www.npr.org/2019/09/26/763545811/how-the-u-s-hacked-isis [Accessed 28 March 2022].

Twitter, 2020. *Civic Integrity: Providing Transparency, Ensuring Credible Information, and Keeping You and Your Data Safe on Twitter*. San Francisco, CA: Twitter. Available from: https://about.twitter.com/en/our-priorities/civic-integrity [Accessed 28 March 2022].

Ukraine Ministry of Defence, 2023. The total combat losses of the enemy from 24.02.2022 to 14.08.2023 [online]. www.mil.gov.ua/en/news/2023/08/12/the-total-combat-losses-of- the-enemy-from-24-02-2022-to-12-08-2023/ [Accessed 14 August 2023].

U.S. Advisory Commission on Public Diplomacy, 2021. *Charter – U.S. Advisory Commission on Public Diplomacy*. Washington, DC: Department of State. Available from: www.state.gov/charter-u-s-advisory-commission-on-public-diplomacy/ [Accessed 28 March 2022].

Walker, S., 2022. Recorded Conversation Between Ukraine's Snake Island Border Guards and Russian Military Ship. *Reddit*, 28 February. Available from: www.reddit.com/r/Milit ary/comments/t0pn7v/recorded_conversation_between_ukraines_snake/ [Accessed 28 March 2022].

Wright, T., 2022. Ukrainian Government Launches Crypto Donation Website with FTX, Kuna and Everstake. *Cointelegraph*, 14 March. Available from: https://cointelegraph. com/news/ukrainian-government-launches-crypto-donation-website-with-ftx-kuna-and-everstake [Accessed 28 March 2022].

Zelenskyy, V., 2022. Speeches. *Government of Ukraine: President of Ukraine*. Available from: www.president.gov.ua/en/news/speeches [Accessed 28 March 2022].

14 Enduring Challenges in Cybersecurity

Responding Quickly and Credibly to Asymmetric Threats

Michael Poznansky

Introduction

A series of significant cyber incidents in recent years has led to spirited debated about how the United States can best defend itself and advance its interests in cyberspace. One such incident was the SolarWinds hack, first made public in December 2020. At that time, the cybersecurity firm FireEye released a report describing a campaign wherein malicious actors compromised software updates for widely used management and accounting software known as SolarWinds Orion. The campaign, which FireEye designated as UNC2452, is thought to have been a Russian-led effort that affected numerous U.S. government agencies, including the Departments of Defense, Commerce, Energy, Homeland Security, Justice, and State (Wilson 2020). Brad Smith, the president of Microsoft, called it "the largest and most sophisticated attack the world has ever seen" (Reuters Staff 2021).

Another significant incident occurred in May 2021. A criminal group known as DarkSide targeted Colonial Pipeline, a major company involved in the provision of oil to the southeastern United States. Using a stolen password, DarkSide was able to launch a ransomware attack. In response, Colonial quietly paid a large sum of money to the criminals and shut down operations to mitigate further damage. The result was panic buying and significant disruption up and down the southern part of the eastern seaboard. This episode held a number of lessons for the government and private sector about potential disruptive actors (it was a criminal group, not a state such as Russia, China, or Iran) and assumptions about the vulnerability of critical infrastructure (Colonial believed the operation of the pipeline and their data systems were isolated from each other) (Sanger and Perlroth 2021).

The Russian invasion of Ukraine, which began on 24 February 2022, has spotlighted some of the unique dynamics of cyber operations in a wartime setting. Some observers expected cyber capabilities to play an important role in the conflict. Jason Healey, for example, argued in late-February 2022, right as the invasion was about to begin that "A Russian invasion of Ukraine may redefine how we think about cyber conflict because it will be the first time a state with real capabilities is willing to take risks and put it all on the line" (Marks 2022). Others have

DOI: 10.4324/9781003425304-15

been more skeptical. The problem, these skeptics argue, is that arguments about cyber redefining warfare fundamentally misunderstand what this tool is good for, namely espionage, disruption, and subversion. It is less effective as a tool of coercion or even a compliment to warfighting (Valeriano et al. 2022). Regardless of its impact, it is clear cyber is playing some role. On the Russian side, reports suggest they used cyber capabilities alongside conventional strikes (Conger and Sanger 2022). On the U.S. side, General Paul Nakasone, the commander of U.S. Cyber Command (USCYBERCOM), confirmed that the United States had "conducted a series of operations across the full spectrum; offensive, defensive, [and] information operations" (Martin 2022).

As these events have been unfolding, senior U.S. policymakers have been working to meet the challenge through a series of institutional changes and policy pronouncements. In May 2021, President Joe Biden released a detailed executive order to "improv[e] the nation's cybersecurity" (The White House 2021). The first head of the recently created Office of the National Cyber Director was confirmed by the Senate on 17 June 2021 (Geller 2021). In early April 2022, the State Department announced the creation of the Bureau of Cyberspace and Digital Policy. Its mission is to "address the national security challenges, economic opportunities, and implications for U.S. values associated with cyberspace, digital technologies, and digital policy" (Department of State 2022). There are clearly more changes on the horizon to the institutional architecture, strategy documents, and policies in the cyber domain.

This chapter explores three challenges facing the United States in cyberspace in the near- to medium-term. The first is the perennial tension between the desire for greater coordination and oversight on the one hand and flexibility, agility, and responsiveness on the other. The second turns on a particular kind of asymmetry in which the United States has certain vulnerabilities that its chief rivals do not, and the effect this has on interactions in cyberspace. The third has to do with the perennial problem of credibility in cyberspace with an eye toward the prospect of achieving deterrence and defending critical infrastructure. While this list is by no means exhaustive, it touches on some of the most important and enduring issues within this domain for the foreseeable future.

Balancing Agility and Coordination

One of the most pressing issues to grapple with in the coming years is how to strike the right balance between responding expeditiously to malicious activity in cyberspace while simultaneously ensuring proper coordination across the federal government (Jensen and Work 2018; SASC 2018). Perhaps unsurprisingly, there has been some back and forth on this issue across administrations.

Former President Donald Trump's widely reported decision in 2018 to give certain entities, most notably USCYBERCOM, greater authority to carry out offensive cyber operations tipped the scales in favor of speed and efficiency (Corn 2021; Geller 2018; Nakashima 2018; Sanger 2018). According to news outlets, by rescinding an Obama-era policy that required interagency coordination of offensive cyber operations, the Trump administration sought to give USCYBERCOM the ability to swiftly take the fight to the adversary without getting bogged down in bureaucratic red tape (Borghard 2018; Volz 2018). The replacement for the Obama directive was widely reported as an "offensive step forward" (Volz 2018). Reports also suggest that the CIA was given more freedom of action in cyberspace under Trump (Dorfman et al. 2020).

When President Biden first took office in January 2021, news outlets indicated that he had kept the basic thrust of Trump's revisions to cyber policy in place, at least for operations of a certain size (Sanger et al. 2021). More recently, however, it was reported that the scales may be shifting a bit. In May 2022, *The Washington Post* published a story detailing some of the changes that were being made by the Biden administration. The overarching policy change from the Trump era which gave USCYBERCOM more flexibility than they had previously was seemingly kept in place. Part of what purportedly changed was that the White House and the State Department would have greater visibility into what was going on in cyberspace (Nakashima 2022).

The benefits of providing greater flexibility and responsiveness to cyber operators are straightforward. Unlike more conventional domains, cyberspace is characterized by constant contact (Fischerkeller and Harknett 2018). As such, the United States must be in a position where it is operating continuously rather than reactively (Fischerkeller and Harknett 2019). Continuous competition is the logic behind strategic concepts such as persistent engagement and defend forward (Kollars and Schneider 2018). Bolstering USCYBERCOM's autonomy was likely part and parcel of such a strategy (Healey 2018). It is also likely why Biden, while making some changes to the architecture governing cyber operations, has not gone as far in the direction of lengthy deliberation as some feared. Nevertheless, there are a series of complex questions involved in this debate that are worth interrogating (Lin 2022).

One commonly discussed risk of granting USCYBERCOM broader authority to act first and unilaterally in cyberspace is that it could inadvertently jeopardize ongoing intelligence operations (Chesney 2018). In this view, the military may choose to conduct an offensive cyber operation against a given target without regard for, or possibly even awareness of, whether U.S. intelligence agencies are currently collecting against that same target. The fact that the commander of USCYBERCOM also serves as the head of the National Security Agency, what is known as a dual-hat role, may mitigate this problem somewhat, but not entirely (Borghard 2018).

Prioritizing speed and efficiency over coordination has several other potential implications. First, it could impact the dynamics of escalation in cyberspace (Valeriano and Jensen 2019). It may be true, as some argue, that the risks are actually negligible given the "self-dampening mechanisms" of cyber operations (e.g., attribution is not instantaneous, it takes time for victims to mount an appropriate counter-response, and there are limits on "the scale and magnitude of the costs that can be imposed solely through cyber campaigns") (Borghard 2018).

While these arguments may apply in the short term, they could be less relevant for escalation risks in the longer term. Even if adversaries do not have the capacity to immediately respond to an increase in the amount and kind of offensive cyber operations aimed at them, they may still seek to invest in new capabilities that could harm the United States in the future. Moreover, even the short-term risks of escalation may be greater than they seem at first. Targets may have prepositioned cyber assets that can be used as retaliation, but which are unknown to the U.S. government – like the recent discovery of Chinese malware in telecommunications systems, electric and gas utilities, and maritime operations and transportation systems in Guam and elsewhere in the United States (Sanger 2023). The states most likely to be potential targets of an increasingly empowered USCYBERCOM – Russia, China, Iran, and North Korea – may be precisely those who have offensive capabilities they can leverage (Healey and Jervis 2020).

This leads naturally to another point, namely that the empowerment of USCYBERCOM and the potential disconnect from other arms of the U.S. government may impede the ability to gain an accurate understanding of who is doing what to whom and why (Smith and Monken 2021). This is especially important for those tasked with defending the nation. If the Department of Homeland Security, the Federal Bureau of Investigation, or even the National Cyber Director – whose job is to "lead the implementation of national cyber policy and strategy" – are not fully apprised of what is happening on the offensive side, they may be caught off guard and less prepared to optimize America's defenses (Costello and Montgomery 2021). Building relationships for coordination purposes can ameliorate this problem to some degree, but there is always a risk of leaving someone in the dark.

This issue is related, but slightly distinct from the challenges of overclassification. According to Healey and Jervis (2021), for example, the layers of secrecy in cyberspace make it difficult to figure out cause and effect. The claim being made here is different but complementary. In short, it is the fact that one arm of the U.S. government may be conducting operations against rivals that lead to a counter-response without the other arms of government being aware of this, and therefore drawing incorrect inferences and failing to anticipate potential retaliation from a defensive standpoint. In this scenario, it is not classification per se that is the problem but rather a lack of coordination.

Moreover, given the breadth of vulnerabilities in cyberspace that U.S. adversaries can exploit, there is a necessary and symbiotic relationship between offensive and defensive operations (Smith and Monken 2021). Offensive tactics inform defensive tactics and vice versa. Reduced visibility among those charged with defending the nation and those carrying out offensive operations against adversaries, while perhaps appealing from the standpoint of maximum efficiency, could inadvertently make the United States less effective at both.

Beyond escalation and coordination, the decision to remove certain constraints on USCYBERCOM could also impact U.S. diplomacy, both the standard kind and the coercive variant. With respect to traditional diplomacy, it is useful for American diplomats and negotiators to understand any potential ongoing operations against the country they are dealing with. As Healey (2018) has argued previously,

> If you're meeting President Xi or Chancellor Merkel, it is not unfair for your NSC to know what U.S. Cyber Command is up to and develop options to slow down (or speed up) such operations to send diplomatic signals or reducing the chances of a mistake which weakens your negotiating position.

Regarding coercive diplomacy, prioritizing speed and efficiency when it comes to carrying out offensive cyber operations could have mixed effects. On the one hand, giving entities like USCYBERCOM more leeway may make it easier to credibly impose costs on rivals, thereby contributing to deterrence; this is part of the argument by proponents of this strategy (see Healey 2019). But if offensive cyber operations are too disconnected from other tools of U.S. statecraft, it may inhibit the ability of decision makers to bring to bear all relevant pressure points for a more holistic coercive strategy (e.g., the imposition of sanctions, and so forth) (Jensen and Work 2018). Additionally, successful coercion in many cases benefit from a degree of reassurance. That is, targets should believe that if they comply with demands they may not only avoid punishment but reap rewards. But if targets come to expect that USCYBERCOM is acting independently of entities that provide these benefits, reassurance is harder.

The foregoing has significant import for the concept of integrated deterrence, which is central to the Department of Defense's strategy and plans under the Biden administration. According to Under Secretary of Defense for Policy Colin Kahl, the idea of integration in this context applies "across domains, so conventional, nuclear, cyber, space, informational" (Garamone 2021). Cyber operations that are carried out independently of other entities responsible for contributing to integrated deterrence against rivals may work at cross-purposes with this concept. Indeed, that may help explain some of the aforementioned changes the Biden administration made to cyber authorities.

Another potential implication of enabling USCYBERCOM to carry out offensive cyber operations without broader input is that the United States could end up in a situation where tactics are driving strategy rather than the reverse (Poznansky 2021). It may well be that in cyberspace, this is inevitable. The fast-moving nature of the domain combined with the reality of constant contact might mean that the best we can ever do is disrupt and degrade the ability of adversaries to do us harm. But, cyber activity does not occur in a vacuum. It is, or at least should be, tied to a state's broader geopolitical objectives. Without conscious deliberation about how any given operation serves the United States' broader foreign policy goals, there is a chance of conducting operations in adversary networks simply because the United States can, without asking whether and under what conditions it should. Many of the current debates about how we ought to conceive of cyberspace – for example, as an intelligence contest (Rovner 2020) or a variant of counterinsurgency (Schroeder et al. 2021) – entail different solutions that may or may not be well served by defend forward and persistent engagement.

To be sure, it may well be that the Biden administration's decision as reported in the *New York Times* to continue providing USCYBERCOM a longer leash to carry out "day-to-day, short-of-war skirmishes in cyberspace" while requiring greater coordination with the National Security Council on larger operations can mitigate many of these challenges (Sanger et al. 2021). But mounting pressure to respond more quickly and forcefully to the spate of recent intrusions could conceivably change things. Moreover, one can easily imagine semantic battles over what constitutes a "significant" attack such that it would require deliberation, or not.

Asymmetric Vulnerabilities

Another issue scholars and practitioners working on cyber issues will have to wrestle with in the coming years turns on asymmetries of various kinds (Grzegorzewski and Marsh 2021). Oftentimes, when the word "asymmetry" is used in the context of cyberspace operations, it is referring to instances in which actors engage in activities below the level of armed conflict to achieve some political objective (Fiala and Worrall 2021); this also sometimes referred to as hybrid warfare and (Kofman 2016) gray-zone conflict (Morris et al. 2019). In the context of this chapter, the term asymmetry refers specifically to the unique set of vulnerabilities the United States has that rivals may not, and its effect on strategic dynamics in cyberspace.

The issue of election meddling specifically, and disinformation more broadly, is emblematic of this problem (Maddox et al. 2021). Figuring out how to guard against malign foreign activity on these fronts is not simply a matter of breaking the code on how to credibly threaten punishment using cyber tools or any other means for that matter. The problem facing the United States is more complicated. One of the core challenges is that the actors most responsible for these activities are not vulnerable in the same way.

Consider that the two main perpetrators of meddling in the 2020 presidential election according to a recently declassified report – Russia and Iran – do not hold competitive, free, and fair elections themselves (National Intelligence Council 2021). Moreover, these states, as well as China, tightly control the Internet (Bandurski 2020). While they are not immune from disinformation, they are likely less vulnerable. Hence the asymmetries. America's rivals are aware of this situation and act accordingly. As Fabian and Berzins (2021) write, Russia subscribes to "the idea that democratic societies are vulnerable to political manipulation." Thomas Rid (2020, p. 11) similarly argues that "disinformation operations, in essence, erode the very foundation of open societies." Foreign actors are thus eager to continue meddling in U.S. elections and propagating disinformation despite attempts to expose and disrupt their ability to do so.

This asymmetry also makes it more difficult to figure out what a proper response should be (bracketing the obvious, bolstering defenses, which should be done regardless). The astute reader may wonder why the United States cannot simply do to rivals what they are doing to the United States. As Former Secretary of Defense Robert Gates (2020, p. 128) recently put it, the United States "also needs to take the offensive from time to time, especially against its primary adversaries. Authoritarian governments must get a taste of their own medicine." This could include carrying out cyber intrusions aimed at delegitimizing and undermining those responsible for malign activity or "interfering in the systems that authoritarian countries use to surveil their own populations" (Buchanan 2021).

U.S. policymakers can obviously try to do these things, but the implications of these actions may not be the same. Research in political science about the different modes of exit – how leaders leave office – in democratic versus authoritarian regimes is relevant here. In democracies, the losers of elections can usually carry on doing whatever it is they wish to do (Chiozza and Goemans 2011). In dictatorships, leaders who lose power face the prospect of exile, punishment, or even death. To put it more concretely, when an adversary interferes in U.S. elections to hurt a candidate and that candidate loses, the consequences are not as dire relative to the costs Putin, Xi, or the Ayatollahs in Iran would face were the United States to stir up the opposition by spreading propaganda and disinformation to undermine their regimes.

This does not mean that interference in elections should be tolerated or embraced; it should not. But it does, unfortunately, make the problem of how to respond more complex. Were the United States to adopt an eye-for-an-eye approach, it may be inherently more escalatory owing to the nature of the target, to say nothing of whether it would be in the United States' interest to go down this road in the first place. The broader point is that the difficulty of threatening retaliation of a similar nature means that policymakers are often left with the choice of imposing costs that are disproportionate – in the direction of either too much, relative to the offense, or too little. This is not necessarily a problem, but it raises the question of what ought to be done when doing too little is unlikely to have a discernible impact on rival behavior and doing too much can heighten tensions.

The Perennial Problem of Credibility

The final issue discussed here has to do with the issue of credibility. When it comes to cyberspace, this issue comes in many forms. One is how to credibly deter malicious attacks. While this is not a new problem, it is worth revisiting some of the key dynamics surrounding it considering developments in recent years as well as identifying under-appreciated challenges.

Early debates about the prospect of deterrence in cyberspace were pessimistic. A big reason for this had to do with the attribution problem (Rid and Buchanan 2015). The basic argument was that if defenders cannot attribute an attack with a high enough degree of certainty, they may be more reticent to respond for fear of imposing significant costs on the wrong actor. And if would-be attackers anticipated this, they may be less inhibited from carrying out cyber-attacks in the first place and dismiss threats to do so as incredible (Nye 2017, pp. 49–50). Former Deputy Secretary of Defense William Lynn put the issue succinctly in 2010: "Whereas a missile comes with a return address, a computer virus generally does not. The forensic work necessary to identify an attacker may take months, if identification is possible at all" (quoted in Nye 2017, p. 50). Part of the difficulty with this argument is that attribution has been rapidly improving in recent years even though it is still far from perfect (Egloff and Smeets 2021, pp. 1–2). And yet, the scope and scale of attacks have continued to grow. There must be additional factors, then, which make credible deterrence challenging.

One is the challenge of devising an appropriate response that will be believed by would-be attackers *and* will be painful enough relative to the benefits they would gain from carrying out cyber intrusions. This is not an easy formula because the two often work at cross-purposes. The most painful responses to cyber-attacks, such as a kinetic military response, are also the least likely to be taken seriously as threats. In most cases, defenders will likely be reluctant to engage in kinetic retaliatory action in response to a cyber-attack for a host of reasons, including the fear of escalation, norms of proportionality, and so forth. The exception, such as in cases of a particularly severe cyber-attack that shuts down part of a country's power grid, is the exception that proves the rule (Zenko 2011).

Conversely, the threats attackers are most likely to believe a victim will follow through on are also the least likely to change their calculus. Consider the threat of economic sanctions. Sanctions have been a popular tool when responding to cyber incidents (Akoto 2021, p. 1084; Buchanan and Sulmeyer 2016, p. 5). President Obama issued several executive orders throughout his presidency giving the Treasury Department sweeping authority to impose economic sanctions on actors who carry out or abet malicious cyber-attacks.[1] Despite its popularity, however, we are lacking reliable evidence of how effective they have been in deterring cyber intrusions. Factors such as the severity of the costs of sanctions on targets and whether international institutions may play a role (Bapat et al. 2013).

Defenders may also threaten to respond to cyber intrusions by carrying out cyber operations of their own. Like sanctions, this may be a mixed bag. To begin with, there is the familiar impediment associated with telegraphing precisely what type of cyber operations would be carried out for fear that doing so would provide the would-be attacker an opportunity to preemptively mitigate vulnerabilities (Gartzke 2013). Threatening to retaliate more generally, perhaps following a demonstration of cyber capabilities in another context to cultivate a reputation for cyber power, may be believed (Poznansky and Perkoski 2018). But whether these threats are worrisome enough to deter malicious behavior is another matter worthy of additional investigation.

A related issue pertaining to credibility and deterrence in cyberspace is how and whether states can set clear boundaries that adversaries believe, if crossed, will lead to severe consequences. In an ideal world, states might be able to increase the credibility of threats of retaliation of various kinds by carving out specific sectors of society or government that are so strategically or politically important they would be forced to respond. While this might inadvertently encourage malicious activity in areas outside of those boundaries, the idea is that by narrowing the scope in this way, leaders are making it more credible that anything is on the table for activity that takes place against specified targets.

Critical infrastructure represents a good example of how this might work in practice – as well as the challenges involved. The United States government has made a significant effort to designate critical infrastructure sectors "whose assets, systems, and networks, whether physical or virtual, are considered so vital to the United States that their incapacitation or destruction would have a debilitating effect on security, national economic security, national public health or safety, or any combination thereof."[2] There are 16 in total. While President George W. Bush first spearheaded the notion of critical infrastructure protection in 2003, the emphasis then was on terrorist attacks. President Obama expanded the aperture to include cyber-attacks in 2013 when he released Presidential Policy Directive 21

(The White House 2013). The issue continued to grow in salience after Russia's meddling in the 2016 presidential elections. President Biden publicly argued that he "will continue to use every tool to deter, disrupt, and if necessary, respond to cyberattacks against critical infrastructure" (Biden 2022).

A potential drawback of defining critical infrastructure too broadly is that it sets up a challenge for those tasked with defending and deterring in cyberspace. By combining such a wide range of sectors under one umbrella, failure to meaningfully respond to attacks in one sector may call into question the credibility that the U.S. government would respond in another, even if the former was less disruptive to U.S. society than the latter (Wilner 2017, p. 314). A similar logic would hold if leaders responded with a less intense form of retaliation for intrusions in one sector than they would have for intrusions against another more important one.

An alternative way of looking at the problem is that defining critical infrastructure so broadly may actually encourage policymakers to impose harsh penalties for attacks against sectors that may be easier to defend or at least get back online than others – precisely for the reason that responding to any attacks is viewed as essential for demonstrating credibility in all sectors – thereby risking escalation in cases when it may be unwise to do so. The recommendation of former high-ranking officials to outline "a clear declaratory policy to respond to malicious activity with both cyber and non-cyber offensive tools" would not resolve this broader issue on its own (Hayden et al. 2021). Rather, an essential part of the conversation would entail difficult discussions about whether to narrow the scope of critical infrastructure sectors the U.S. was willing to impose harsh penalties for targeting or, at the very least, creating some sort of hierarchy of importance and making this widely known.

The Road Ahead

The challenges identified here are a fraction of those confronting the United States. Others include whether the scale and scope of intrusions like SolarWinds render them distinct from traditional espionage, how to grapple with the fact that many of our rivals do not draw the same lines between national security espionage and economic espionage as the United States does, and how the pervasive secrecy that characterizes much of what goes on in cyberspace impacts the ability to develop norms and robust public-private partnerships. Nevertheless, being clear-eyed about the pros and cons of greater coordination versus greater latitude and having greater awareness of America's unique vulnerabilities and the continued challenges of credibility – the issues dealt with here – would put the United States in a better position in the coming years.

Notes

1 See, for instance, https://home.treasury.gov/policy-issues/financial-sanctions/faqs/topic/1546

2 See www.cisa.gov/critical-infrastructure-sectors

References

Akoto, W., 2021. International Trade and Cyber Conflict: Decomposing the Effect of Trade on State-Sponsored Cyber Attacks. *Journal of Peace Research*, 58(5), 1083–1097.

Bandurski, D., 2020. A Brief Experiment in a More Open Chinese Web. *Brookings Institution*, 12 November. Available from: www.brookings.edu/articles/a-brief-experiment-in-a-more-open-chinese-web/

Bapat, N. A., Heinrich, T., Kobayashi, Y., and Clifton Morgan, T., 2013. Determinants of Sanctions Effectiveness: Sensitivity Analysis Using New Data. *International Interactions*, 39(1): 79–98. https://doi.org/10.1080/03050629.2013.751298.

Biden, J. R., 2022. Statement by President Biden on Our Nation's Cybersecurity. *The White House Briefing Room*, 21 March. Available from: www.whitehouse.gov/briefing-room/statements-releases/2022/03/21/statement-by-president-biden-on-our-nations-cybersecurity/

Borghard, E. D., 2018. What Do the Trump Administration's Changes to PPD-20 Mean for U.S. Offensive Cyber Operations? *Net Politics*, 10 September. Available from: www.cfr.org/blog/what-do-trump-administrations-changes-ppd-20-mean-us-offensive-cyber-operations

Buchanan, B., 2021. The SolarWinds Hack Is Just the Beginning: The United States Must Learn to Live With Cyber-Espionage. *Foreign Affairs*, 16 April. Available from: www.foreignaffairs.com/articles/united-states/2021-04-16/solarwinds-hack-just-beginning

Buchanan, B. and Sulmeyer, M., 2016. Russia and Cyber Operations: Challenges and Opportunities for the Next U.S. Administration. *Carnegie Endowment for International Peace and The Chicago Council on Global Affairs: Task Force on U.S. Policy Toward Russia, Ukraine, and Eurasia*, 1–6. Available from: https://carnegieendowment.org/2016/12/13/russia-and-cyber-operations-challenges-and-opportunities-for-next-u.s.-administration-pub-66433

Chesney, R., 2018. Offensive Cyber Operations and the Interagency Process: What's at Stake With the New Trump Policy. *Lawfare*, 16 August. Available from: www.lawfaremedia.org/article/offensive-cyber-operations-and-interagency-process-whats-stake-new-trump-policy

Chiozza, G., and Goemans, H. E., 2011. *Leaders and International Conflict*. Cambridge: Cambridge University Press.

Conger, K., and Sanger, D. E., 2022. Russia Uses Cyberattacks in Ukraine to Support Military Strikes, Report Finds. *New York Times*, 2 May. Available from: www.nytimes.com/2022/04/27/us/politics/russia-cyberattacks-ukraine.html

Corn, G., 2021. Authorities and Legal Considerations for US Cyber Command and Information Operations in a Contested Environment. *Modern War Institute*, 29 March. Available from: https://mwi.westpoint.edu/authorities-and-legal-considerations-for-us-cyber-and-information-operations-in-a-contested-environment/

Costello, J. and Montgomery, M., 2021. How the National Cyber Director Position Is Going to Work: Frequently Asked Questions. *Lawfare*, 24 February. Available from: www.lawfareblog.com/how-national-cyber-director-position-going-work-frequently-asked-questions

Department of State, 2022. Establishment of the Bureau of Cyberspace and Digital Policy. *Office of the Spokesperson*, 4 April. Available from: www.state.gov/establishment-of-the-bureau-of-cyberspace-and-digital-policy/

Dorfman, Z., Zetter, K., McLaughlin, J., and Naylor, S. D., 2020. Exclusive: Secret Trump Order Gives CIA More Powers to Launch Cyberattacks. *Yahoo! News*, 15 July. Available from: https://news.yahoo.com/secret-trump-order-gives-cia-more-powers-to-launch-cyberattacks-090015219.html

Egloff, F. J. and Smeets, M., 2021. Publicly Attributing Cyber Attacks: A Framework. *Journal of Strategic Studies*, 46(3), 502–533. https://doi.org/10.1080/01402 390.2021.1895117.

Fabian, S. and Berzins, J., 2021. Striking the Right Balance: How Russian Misinformation Operations in the Baltic States Should Inform US Strategy in Great Power Competition. *Modern War Institute*, 12 April. Available from: https://mwi.usma.edu/striking-the-right-balance-how-russian-information-operations-in-the-baltic-states-should-inform-us-strat egy-in-great-power-competition/

Fiala, O. C. and Worrall, J., 2021. Imposing Costs: Unconventional Warfare in the Information Environment. *Modern War Institute*, 6 July. Available from: https://mwi.usma.edu/impos ing-costs-unconventional-warfare-in-the-information-environment/

Fischerkeller, M. and Harknett, R. J., 2018. Persistent Engagement and Tacit Bargaining: A Path Toward Constructing Norms in Cyberspace. *Lawfare*, 9 November. Available at: www. lawfareblog.com/persistent-engagement-and-tacit-bargaining-path-toward-constructing-norms-cyberspace

Fischerkeller, M. and Harknett, R. J., 2019. Persistent Engagement, Agreed Competition, and Cyberspace Interaction Dynamics and Escalation. *The Cyber Defense Review*, 267–287. Available at: https://cyberdefensereview.army.mil/Portals/6/CDR-SE_S5-P3-Fische rkeller.pdf

Garamone, J., 2021. Concept of Integrated Deterrence Will Be Key to National Defense Strategy, DOD Official Says. *DOD News*, 8 December. Available at: www.defense.gov/ News/News-Stories/Article/Article/2866963/concept-of-integrated-deterrence-will-be-key-to-national-defense-strategy-dod-o/

Gartzke, E., 2013. The Myth of Cyberwar: Bringing War in Cyberspace Back Down to Earth. *International Security*, 38(2): 41–73. https://doi.org/10.1162/ISEC_a_00136.

Gates, R. M., 2020. The Overmilitarization of American Foreign Policy: The United States Must Recover the Full Range of Its Power. *Foreign Affairs*, 99(4), 121–132.

Geller, E., 2018. Trump Scraps Obama Rules on Cyberattacks, Giving Military Freer Hand. *Politico*, 16 August. Available from: www.politico.com/story/2018/08/16/trump-cybers ecurity-cyberattack-hacking-military-742095

Geller, E., 2021. Senate Confirms Chris Inglis as Biden's Top Cyber Adviser. *Politico*, 17 June. Available from: www.politico.com/news/2021/06/17/senate-confirms-chris-inglis-cyber-495075

Grzegorzewski, M. and Marsh, C., 2021. Incorporating the Cyberspace Domain: How Russia and China Exploit Asymmetric Advantages in Great Power Competition. *Modern War Institute*, 15 March. Available from: https://mwi.usma.edu/incorporating-the-cybersp ace-domain-how-russia-and-china-exploit-asymmetric-advantages-in-great-power-comp etition/

Hayden, M., Ridge, T., Shkor, J. and Montgomery, M., 2021. A Strong Offense Can Decrease Cyberattacks on Critical Infrastructure. *The Hill*, 19 February. Available from: https:// thehill.com/opinion/cybersecurity/539085-a-strong-offense-can-decrease-cyber-attacks-on-critical-infrastructure/

Healey, J., 2018. Memo to POTUS: Responding to Cyber Attacks and PPD-20. *The Cipher Brief*, 24 May. Available from: www.thecipherbrief.com/column_article/memo-potus-res ponding-cyber-attacks-ppd-20

Healey, J., 2019. The Implications of Persistent (and Permanent) Engagement in Cyberspace. *Journal of Cybersecurity*, 5(1): 1–15. https://doi.org/10.1093/cybsec/tyz008.

Healey, J. and Jervis, R., 2020. The Escalation Inversion and Other Oddities of Situational Cyber Stability. *Texas National Security Review*, 3(4): 30–53. http://dx.doi.org/10.26153/tsw/10962.

Healey, J. and Jervis, R., 2021. Overclassification and Its Impact on Cyber Conflict and Democracy. *Modern War Institute*, 22 March. Available from: https://mwi.usma.edu/overclassification-and-its-impact-on-cyber-conflict-and-democracy/

Jensen, B. and Work, JD., 2018. Cyber Civil-Military Relations: Balancing Interests on The Digital Frontier. *War on the Rocks*, 4 September. Available from: https://warontherocks.com/2018/09/cyber-civil-military-relations-balancing-interests-on-the-digital-frontier/

Kofman, M., 2016. Russian Hybrid Warfare and Other Dark Arts. *War on the Rocks*, 11 March. Available from: https://warontherocks.com/2016/03/russian-hybrid-warfare-and-other-dark-arts/

Kollars, N. and Schneider, J., 2018. Defending Forward: The 2018 Cyber Strategy Is Here. *War on the Rocks*, 20 September. Available from: https://warontherocks.com/2018/09/defending-forward-the-2018-cyber-strategy-is-here/

Lin, H., 2022. President Biden's Policy Changes for Offensive Cyber Operations. *Lawfare*, 17 May. Available from: www.lawfareblog.com/president-bidens-policy-changes-offensive-cyber-operations

Maddox, JD, Gentzel, C. and Levis, A., 2021. Toward a Whole-of-Society Framework for Countering Disinformation. *Modern War Institute*, 10 May. Available from: https://mwi.usma.edu/toward-a-whole-of-society-framework-for-countering-disinformation/

Marks, J., 2022. Here's What Cyber Pros Are Watching in the Ukraine Conflict. *The Washington Post*, 14 February. Available from: www.washingtonpost.com/politics/2022/02/24/heres-what-cyber-pros-are-watching-ukraine-conflict/

Martin, A., 2022. US Military Hackers Conducting Offensive Operations in Support of Ukraine, Says Head of Cyber Command. *Sky News*, 1 June. Available from: https://news.sky.com/story/us-military-hackers-conducting-offensive-operations-in-support-of-ukraine-says-head-of-cyber-command-12625139

Morris, L. J., et al. 2019. *Gaining Competitive Advantage in the Gray Zone*. Santa Monica, CA: RAND Corporation.

Nakashima, E., 2018. Trump Gives the Military More Latitude to Use Offensive Cyber Tools Against Adversaries. *The Washington Post*, 16 August. Available from: www.washingtonpost.com/world/national-security/trump-gives-the-military-more-latitude-to-use-offensive-cyber-tools-against-adversaries/2018/08/16/75f7a100-a160-11e8-8e87-c869fe70a721_story.html

Nakashima, E., 2022. The Biden Administration Is Refining a Trump-Era Cyber Order. *The Washington Post*, 13 May. Available from: www.washingtonpost.com/politics/2022/05/13/biden-administration-is-refining-trump-era-cyber-order/

National Intelligence Council, 2021. *Foreign Threats to the 2020 US Federal Elections*. Washington, DC: National Intelligence Council.

Nye, J. S., 2017. Deterrence and Dissuasion in Cyberspace. *International Security*, 41(3), 44–71. https://doi.org/10.1162/ISEC_a_00266.

Poznansky, M., 2021. Covert Action, Espionage, and the Intelligence Contest in Cyberspace. *War on the Rocks*, 23 March. Available from: https://warontherocks.com/2021/03/covert-action-espionage-and-the-intelligence-contest-in-cyberspace/

Poznansky, M. and Perkoski, E., 2018. Rethinking Secrecy in Cyberspace: The Politics of Voluntary Attribution. *Journal of Global Security Studies*, 3(4), 402–416. https://doi.org/10.1093/jogss/ogy022.

Reuters Staff, 2021. SolarWinds Hack Was "Largest and Most Sophisticated Attack" Ever: Microsoft President. *Reuters*, 14 February. Available from: www.reuters.com/arti cle/us-cyber-solarwinds-microsoft-idUSKBN2AF03R

Rid, T. and Buchanan, B., 2015. Attributing Cyber Attacks. *Journal of Strategic Studies*, 38(1-2), 4–37. DOI: 10.1080/01402390.2014.977382

Rid, T., 2020. *Active Measures: The Secret History of Disinformation and Political Warfare*. New York: Farrar, Strauss and Giroux.

Rovner, J., 2020. What Is an Intelligence Contest? *Texas National Security Review*, 3(4), 114–120. Available from: https://repositories.lib.utexas.edu/bitstream/handle/2152/83955/TNSRVol3Iss4Rovner.pdf

Sanger, D. E., 2018. Trump Loosens Secretive Restraints on Ordering Cyberattacks. *New York Times*, 20 September. Available from: www.nytimes.com/2018/09/20/polit ics/trump-cyberattacks-orders.html

Sanger, D. E., 2023. Chinese Malware Hits Systems on Guam. Is Taiwan the Real Target? *New York Times*, 24 May. Available from: www.nytimes.com/2023/05/24/us/politics/china-guam-malware-cyber-microsoft.html

Sanger, D. E., Barnes, J. E. and Perlroth, N., 2021. Preparing for Retaliation Against Russia, U.S. Confronts Hacking by China. *New York Times*, 25 October. Available from: www.nytimes.com/2021/03/07/us/politics/microsoft-solarwinds-hack-russia-china.html

Sanger, D. E. and Perlroth, N., 2021. Pipeline Attack Yields Urgent Lessons About U.S. Cybersecurity. *New York Times*, 14 May. Available from: www.nytimes.com/2021/05/14/us/politics/pipeline-hack.html

SASC. 2018. *Stenographic Transcript Before the Committee on Armed Services, United States Senate: United States Cyber Command, Feb. 27, 2018*. Washington, DC.

Schroeder, E., Handler, S. and Herr, T. 2021. Population-Centric Cybersecurity: Lessons From Counterinsurgency. *Modern War Institute*, 17 May. Available from: https://mwi.usma.edu/population-centric-cybersecurity-lessons-from-counterinsurgency/

Smith, M. and Monken, J., 2021. The Colonial Pipeline Hack Shows We Need a Better Federal Cybersecurity Ecosystem. *Modern War Institute*, 1 June. Available from: https://mwi.usma.edu/the-colonial-pipeline-hack-shows-we-need-a-better-federal-cybersecur ity-ecosystem/

The White House, 2013. Presidential Policy Directive – Critical Infrastructure Security and Resilience. *Office of the Press Secretary*, 12 February. Available from: https://obamawhi tehouse.archives.gov/the-press-office/2013/02/12/presidential-policy-directive-critical-infrastructure-security-and-resil

The White House, 2021. Executive Order on Improving the Nation's Cybersecurity. *Presidential Actions*, 12 May. Available from: www.whitehouse.gov/briefing-room/presi dential-actions/2021/05/12/executive-order-on-improving-the-nations-cybersecurity/

Valeriano, B. and Jensen, B., 2019. The Myth of the Cyber Offense: The Case for Restraint. *CATO Institute, Policy Analysis No. 862*. Available from: www.cato.org/policy-analysis/myth-cyber-offense-case-restraint

Valeriano, B., Lonergan, E. D., Lonergan, S. W. and Jensen, B., 2022. Putin's Invasion of Ukraine Didn't Rely on Cyberwarfare. Here's Why. *The Washington Post*, 7 March. Available from: www.washingtonpost.com/politics/2022/03/07/putins-invasion-ukraine-didnt-rely-cyber-warfare-heres-why/

Volz, D., 2018. Trump, Seeking to Relax Rules on U.S. Cyberattacks, Reverses Obama Directive. *Wall Street Journal*, 15 August. Available from: www.wsj.com/articles/trump-seeking-to-relax-rules-on-u-s-cyberattacks-reverses-obama-directive-1534378721

Wilner, A., 2017. Cyber Deterrence and Critical-Infrastructure Protection: Expectation, Application, and Limitation. *Comparative Strategy*, 36(4), 309–318. https://doi.org/10.1080/01495933.2017.1361202.

Wilson, S., 2020. SolarWinds Recap: All of the Federal Agencies Caught Up in the Orion Breach. *FedScoop*, 22 December. Available from: https://fedscoop.com/solarwinds-recap-federal-agencies-caught-orion-breach/

Zenko, M., 2011. Cyber Attacks and Military Responses. *CFR*, 3 June. Available from: www.cfr.org/blog/cyber-attacks-and-military-responses

Index

asymmetric vulnerabilities 204–5

Baltics: resilience in information environment 159–60; and Russia's information and cyber operations in 52–3, 158–9; *see also* Russia

China: concept of information operations 165–6; cyber operations 13–17; cyber attacks against the United States 164–5; information operations in Africa 88–90; information operations in Latin America 106–8; information operations against Taiwan 161–3; military expansion of 91; and People's Liberation Army 161–2
colonial pipeline 145, 148; *see also* Ransomware
competition: in cyber conflict 120–2; the essence of 27–31; and Great Power Competition literature 32–7; *see also* Great Power Competition
contactless War 46; *see also* New Generation Warfare
critical service delivery in 148–9
cyber attacks: Baltic states' resilience against 159–60; history of 128–30; ransomware 142–8; Taiwan's resilience against 163–4; against Ukraine 60–2; against the West 63–4; against the United States 160–1, 164–5
cyber conflict: competition in 120–2; the lower bound of intensity of 117–18; the upper bound of intensity of 116–17; and the way forward in 122–3
cyber crime 145–7; *see also* Ransomware
cyber: massing effects in 135–7; protraction in 115–16, 118–20; and

social network analysis in 131–4; threat landscape in Latin America 101–3
cyber security: limits of government intervention 144–5; management and mitigation within 149–50; measures for military organizations 180; reporting standards within 150–1; system of systems approach 151–3; and USCYBERCOM role in 200–3

deterrence 205–7
digital dust 176–7; *see also* unintentional trace data
digital literacy 186
digital signature management 178–9
DIME 39–40, 104
disinformation: countering of 184–5; disruption of 186–7; regulation of 187–8; resilience against 185–6; and Ukraine's use of social media against 190–1

force protection strategy: compliance auditing 180; digital signature awareness training as 179; and technical systems architecture 179–80
FSB: as an actor in information operations 47–8, 59, 64; and role in cyber attacks 64–5; *see also* information operations

Great Power Competition: bias in literature 32–7; compared to Cold War 37–8; implications of democratized intelligence for 80–2; in Latin America 103–4; as a military struggle 35–7; Special Operations Forces in 174–5; and an unintegrated approach to 34–5; *see also* competition

GRU: in active measures 9, 11–12, 30–1; as an actor in information operations 47–8, 51; and role in cyber attacks 59–62, 64–5; *see also* Information Operations

influence operations 47; *see also* information operations
information operations: in Africa 88–90, 91–5; in the Baltics 52–3, 158–9; Chinese concept of 165–6; imagery-based 191–3; in Latin America 104–8; organizations involved in 47–8; in surveillance economy 178; against Taiwan 161–3; against Ukraine 62–3; Ukraine's whole-of-society approach to 188–90; against the United States 160–1; and United States' definition of 184; *see also* influence operations
intelligence: collection of 76–8; definitions of 75–6; history of 73–5; intelligence analysis 78–80; implications of democratized intelligence for Great Power Competition 80–2; and in surveillance economy 178
Internet Research Agency 7, 40, 59, 63, 67, 94
irregular warfare 134–5, 166–7

Kharkiv News Agency 7

Latin America: Chinese information operations in 106–8; Great Power Competition in 103–4; information operations in 104–8; and threat landscape in 101–3

massing effects 135–7
military expansion 91

NATO: enhanced forward presence 45; cyber defenses of 66–7
New Generation Warfare 44, 46–7

People's Liberation Army 161–2
protraction 115–16
psychological operations (PSYOPS): and cyber space 130–1; definition of 129–30; role in massing effects 135–8; targeting cycle of 133; and social network analysis 132–4

Ransomware: definition and examples of 142–4; realities of 145–7; and responses to 147–8; *see also* cyber attacks
Russia: current cyber capabilities 65–6; cyber operations 8–13; cyber operations in Ukraine 61–3; the concept of cyber operations 49–52, 57–60; information operations in Africa 91–5; information and cyber operations in the Baltics 52–3, 158–9; information operations in Latin America 104–6; information and cyber operations against the United States 160–1; and use of unconventional warfare 158–61
Russian Today (RT): in Africa 93–4; and in Latin America 104–6

Sixth Generation Warfare 46; *see also* New Generation Warfare
social network analysis (SNA) 131–4
special operations forces (SOF): role in cyberspace 174–5
surveillance economy 177–8
SVR: as an actor in information operations 47; *see also* information operations

Taiwan: information operations against 161–3; and resilience against cyber attacks 163–4

ubiquitous technical surveillance 175–6
Ukraine: cyber operations against 60–3; whole-of-society approach to information operations 188–90; and use of social media against 190–1; *see also* cyber attacks
unconventional warfare: Chinese use of 161–5; definition of 157–8; possible defense against 166–8; Russia's use of 158–61; United States' perception of 166–7; *see also* irregular warfare
unintentional trace data 176–7; *see also* digital dust
USCYBERCOM 200–3

Vladimir Slipchenko 46; *see also* New Generation Warfare

Wagner (Private Military Company) 31, 40, 77, 94–5

9781032545295